DINOSAUR
LIVES

ALSO BY JOHN R. HORNER

Digging Dinosaurs
(with James Gorman)
Maia: A Dinosaur Grows Up
(with James Gorman)
The Complete T. Rex
(with Don Lessem)
Digging up Tyrannosaurus Rex
(with Don Lessem)

DINOSAUR LIVES

Unearthing an Evolutionary Saga

JOHN R. HORNER
AND EDWIN DOBB

HarperCollins*Publishers*

DINOSAUR LIVES. Copyright © 1997 by John R. Horner and Edwin Dobb. All rights reserved. Printed in the United States of America. No part of this book may be used or reproduced in any manner whatsoever without written permission except in the case of brief quotations embodied in critical articles and reviews. For information address HarperCollins Publishers, Inc., 10 East 53rd Street, New York, NY 10022.

HarperCollins books may be purchased for educational, business, or sales promotional use. For information please write: Special Markets Department, HarperCollins Publishers, Inc., 10 East 53rd Street, New York, NY 10022.

FIRST EDITION

Designed by Joseph Rutt

Library of Congress Cataloging-in-Publication Data

Horner, John R.
    Dinosaur lives : unearthing an evolutionary saga / by John R. Horner
and Edwin Dobb. — 1st ed.
        p.   cm.
    ISBN 0-06-017486-2
    1. Dinosaurs.   I. Dobb, Edwin, 1950–   .   II. Title.
QE862.D5H644   1997
567.9—dc21   97-8168

97 98 99 00 01 ❖/RRD 10 9 8 7 6 5 4 3 2 1

CONTENTS

ACKNOWLEDGMENTS

Neither this book nor the research it describes would have been possible without the help of a great many people. These include Judy Weaver, Gail Cary, Shelly McKamey, and the entire staff of the Museum of the Rockies. I'm especially grateful to my friends, past and present, in the paleontology department—Pat Leiggi, Bob Harmon, Carrie Ancell, Ellen Lamm, Allison Gentry, Bea Taylor, Frankie Jackson, Jill Peterson, Matt Smith, and Karen Chin.

Undergraduate, graduate, and postgraduate students who contributed hard work, fresh ideas, and significant discoveries include David Varricchio, Yoshi Katsura, Greg Erickson, Mary Schweitzer, Kristi Curry, Rod Scheetz, Rebecca Laws, Pat Druckenmiller, Jill Peterson, April Lafferty, Peter Nassar, Vicki Clouse, Beverly Eschberger, Jody Smith, Des Maxwell, Ray Rogers, David Dilkes, and Scott Sampson.

And where would we professionals be without the help of amateur collectors and volunteer field specialists? I'm thinking now of Ken Olson, Curt Padilla, John Bruninga, Greg Clouse, Gloria Siebrecht, Mark Lindner, Jay Grimaldi, David Smith, Bob Downs, Sid Hoffsteder, Betty Quinn, Pat Murphy, Donna Coles, Gail McCrimmon, Wendy Sloboda, Nate Murphy, Hilory Korte, the many crews of Camp Makela, and especially my son Jason, who always managed to show up his father with better fossils.

For generously allowing us to collect on their land I want to express my gratitude to Chief Earl Oldperson, Don Dubray, and the Blackfeet Nation, the Lewis and Vernon Carroll families of Cut

Bank, Huey Monroe, Ricky Reagan, Gloria Sunquist, Truman Hall, Mark and Traci Tilstra, the McDowell family of Bridger, Marge Baisch and the Eaton Ranch. Logistical help, for which I'm also grateful, came from Chuck McAlpine and Marvin Weatherwax. I owe a debt of thanks as well to the Nature Conservancy, the Montana State Department of Natural Resources and Conservation, the Bureau of Land Management, the Bureau of Reclamation, and the Army Corps of Engineers.

Paleontology receives only a fraction of the financial assistance granted to other scientific fields but that fraction makes all the difference in the world. Our work during the past ten years has been supported by the National Science Foundation, National Geographic, Turner Foundation, Windway Foundation, and the MacArthur Foundation. We also received major donations from Jim and Bea Taylor, Barbara and Robbie Lee, Terry and Mary Kohler, Donna and Michael Coles, Steven Spielberg, and Damaris Waggoner. Thanks to you all.

Thanks also to those who donated equipment, services, and technical advice, including Cemax, Silicon Graphics, Sun Microsystems, Polaroid, Glenn Daleo and San Diego Children's Hospital, Sylvia and Peter Berens of Apunix, Bozeman Deaconess Hospital, and the General Electric Jet Engine Laboratories in Cincinatti, Ohio. And thanks to Canada Fossils Limited.

To colleagues Kevin Padian, Armand de Ricqlès, Phil Currie, Ken Carpenter, the late, great Karl Hirsch, Jamie Powell, Dan Grigorescu, Rinchen Barsbold, John Ostrom, Dave Weishampel, Jim Knight, Ernie Vyse, Michael Ghiselin, Geerat Vermeij, John Ruben, and Jim Schmitt, I'm deeply grateful for countless stimulating conversations regarding dinosaurs, cladistics, and evolution.

My thanks to photographers Bruce Selyem and Terry Panasuk for their memorable pictures of specimens and excavations.

And a very special thanks to my wife Celeste Clair Horner, who did much of the artwork for the book.

Last, to Rob Makela: thanks, friend, for remaining close by. Ed and I are indebted to our agent Kris Dahl, who suggested the collaboration, and to our editor at HarperCollins, Eamon Dolan, who weathered it.

1

DINOSAURS IN CAPTIVITY?

Velociraptor. Translation: quick plunderer. A creature that lived up to its name, and then some. Though no heavier than an adult human, *Velociraptor* was probably pound-for-pound one of the fiercest meat-eating dinosaurs that ever lived. Consider its unusually large eyes and especially the long, clamplike jaws lined with batteries of serrated fangs. No blunt plant-mashing molars, only fangs, each one of them slightly curved and saw-sharp. Walking on its hind legs like *Tyrannosaurus, Velociraptor* also possessed muscular arms and agile hands capable of grasping and crushing. Deadliest of all was the sicklelike claw attached to the second toe of each of its powerful legs.

Velociraptor likely hunted in packs, sometimes attacking dinosaurs several times its size, kicking and tearing deep gashes with its feet and ripping through flesh with its teeth until its quarry collapsed from shock, loss of blood, or exhaustion. Prey often were eaten alive. What this ruthless predator did not do, however, was menace other creatures with a forked tongue that protruded from its mouth. *Velociraptor* may have been terrible, but it was no lizard, a distinction I took pains to make while working as the paleontological consultant on the 1993 film *Jurassic Park*.

On the day I have in mind, Steven Spielberg and his crew had planned to shoot the kitchen chase, a scene, you may recall, that takes place late in the story, after a pair of *Velociraptors* have escaped from their cage and entered the Visitor Center. There the rampaging carnivores chance upon the children, Tim and Lex, whose visit with their grandfather, the park's misguided founder, has taken one horrific turn after another. The kids flee into the spacious cafeteria kitchen and hide at the far end of a stainless steel counter. At that point, according to the script, the raptors were supposed to burst through the kitchen door and, while stomping down the aisle toward Tim and Lex, probe the air with swift, serpentine tongues, further unnerving the children, not to mention the audience.

The problem, as I explained to Spielberg on the set, is that dinosaur tongues were neither forked nor equipped to detect odors. To suggest that they were would reinforce the long-held but mistaken idea that dinosaurs were reptilian—that is, small-brained, sluggish, and cold-blooded. "Whatever you have the raptors do in the kitchen," I advised, "don't break from the pattern we've already established—that they resembled birds."

Spielberg took my criticism in stride, turning the problem over to his crew, who came up with an inspired solution. Remember when the head of the first *Velociraptor* appears in the round window of the kitchen door? While surveying the room with its large right eye, the dinosaur expels a blast of hot air from its nostril, momentarily steaming up the glass. That angry snort is something no cold-blooded reptile could do. Spielberg's technical assistants had devised a frightening visual effect consistent with what we now know about *Velociraptor,* producing one of the most memorable images of the film.

On another occasion the crew tried to coax a life-size model of *Tyrannosaurus* into walking correctly. Actually, the model, though built to full scale, was incomplete; it included only the legs, tail, and lower half of the body. That's because the subject of the shot was supposed to be one of the dinosaur's feet in motion. The large model was driven by internal machinery that was in turn controlled electronically by a technician who operated a much smaller model exactly as if he were a puppeteer. In fact, that's what he was called. But instead of handling strings, this puppeteer worked levers, and whatever motions he induced in his puppet dinosaur, the life-size model mimicked. "How's that look?" he asked me.

Well, I had to admit, the overall setup was pretty impressive. Spielberg, his crew, and I wore raincoats to protect us from artificial rain that fell from the ceiling of the hangarlike studio. Before us loomed half of a very lifelike tyrannosaur that brought its ponderous foot to rest in a shallow pool of mud, lifted it while the crew smoothed over the surface of the pool, then let it drop again. There was just one problem: No tyrannosaur I know would have walked the way this one did. To be specific, as it carried its leg forward the creature placed the front part of its foot down first, toe-to-heel, as a prancing horse might. Dinosaur anatomy simply did not allow for that motion, and if for some reason the animal had tried to perform it, it would probably have fallen over, never to rise again (more about this shortly).

"Here's how it should work," I said, taking control of the levers on the electronic puppet. But translating the movements of the small model into those of the large one turned out to be harder than I expected. Spielberg withdrew to one side and waited more or less patiently while we rehearsed the dinosaur one-step, polishing that single, simple gesture to perfection. After an hour or so, we were ready. All the audience sees of our effort is the foot of a giant tyrannosaur slamming into the rain-soaked ground like a pile driver. Brief though it may be, however, the scene is one of the most ominous in *Jurassic Park*. For me, it is a satisfying one as well, because the movement is anatomically correct—the heel arrives before the toe.

Some of my suggestions made it into the movie but not into movie theaters. I'm thinking now of the scene in which the tyrannosaur attacks the annoying lawyer. To appreciate what I'm about to tell you, you have to know that dinosaur teeth never stopped growing. As the animals' chewing surfaces wore down, new teeth replaced the old. Breaking and losing teeth was commonplace, especially among theropods, the bipedal, meat-eating group to which *Velociraptor* and *Tyrannosaurus* belong, because they tended to rip flesh and sinew with great force. On numerous digs I've uncovered theropod teeth, and nothing but their teeth, mingled with the bones of other dinosaurs that showed unmistakable signs of having been chewed. It would be altogether normal, I told Spielberg, that if the tyrannosaur in *Jurassic Park* got ahold of anything in its jaws—the Jeep it chases or the lawyer it mauls—a tooth would snap off or come loose. Since the story didn't permit the dinosaur to catch up with the Jeep, Spielberg chose the second option, concluding the attack scene with a shot of an eight-inch-long tyrannosaur tooth lodged in the lawyer's leg. That version was accurate, certainly, not to mention riveting, but on reconsideration Spielberg and his colleagues deemed it too gruesome for children. The lost-tooth shot was pulled from the film.

I should mention, of course, that as the scientific adviser to *Jurassic Park* I could do no more than make suggestions, some of which were adopted, many of which were not. The filmmakers took liberties whenever they felt such liberties would improve the story. There is no evidence that *Dilophosaurus* could spit, for instance, much less spit a toxic substance, as it does during the storm, shortly after the park's security systems shut down. Nor can anyone say with confidence that it had a neck frill like the one shown in the movie. *Dilophosaurus* is known only from one badly preserved skeleton and a few fragments, and they tell us little more than that it was a meat eater and as an adult grew to about fifteen feet in length. Another departure from present paleontological evidence involves the aggressive behavior of the rogue tyrannosaur, especially when it chases the Jeep in which Ellie, the biologist, and Ian, the mathematician, are trying to escape. This example is worth exploring in some detail because the filmmakers' depiction reinforces a widespread but discredited idea about this seemingly well-known dinosaur.

First, it's important to keep in mind that only about two dozen *Tyrannosaurus rex* skeletons have been found in the entire world, and only a few of those, including one my staff and I excavated in eastern Montana in 1990, are complete. Second, an improperly reconstructed *T. rex* skeleton stood in the American Museum of Natural History from the early 1900s, shortly after the first such fossils were discovered, also in Montana, until 1992, when it was finally dismantled. For most of this century, in other words, visitors to the most prestigious natural history museum in the country were told in the most memorable terms possible that *Tyrannosaurus rex,* the "tyrant lizard king," was a vicious predator that walked with its tail on the ground, scanning the horizon for its next victim. Not much of this picture agrees with the evidence, however. Like all dinosaurs' tails, the tail of *T. rex* contained tendons that kept it erect. Any specimen display that shows the tail resting on the ground is displaying a broken tail. And to make a dinosaur stand upright while looking straight ahead, the back must be broken as well. So must the neck, in two places. Properly understood, *T. rex*'s skeleton tells us instead that the six-ton animal typically leaned forward, counterbalancing the weight of its immense head and hefty tail over an anatomical pivot—its legs.

That's the evidence that was misinterpreted. Here's the evidence that was overlooked until very recently: *T. rex*'s arms are so short that they cannot be joined together; they cannot grasp themselves, to say nothing of another animal that happens to be running away. What's more, the ratio of *T. rex*'s upper leg bone to its lower strongly suggests that, while it may have been able to lumber pretty fast compared to you and me, the animal was much better suited for long-distance walking. And consider this: that massive head, with jaws that were four feet long and packed with six-inch teeth, contained tiny eye sockets. Tiny sockets mean tiny eyes, and tiny eyes imply poor vision. Finally, there is the matter of the brain case. From the internal shape of the skull we can determine the size of

Tyrannosaurus rex walked on its hind legs using its tail for balance.
T. rex had tiny arms, useless for grasping prey. The center of gravity for
this 12,000-pound dinosaur was twelve feet off the ground, making it
very unstable and an unlikely candidate for rapid, agile moves.

various parts of the brain. *T. rex*'s brain was small—we can be sure
the creature wasn't solving algebraic equations or writing novels—
but the olfactory lobe, where the sense of smell is seated, was disproportionately large. In fact, it was huge. The only creatures alive
today with a comparable olfactory lobe are the kiwi, which is nocturnal and relies on smell to survive, and the vulture, which can
detect the scent of rotting flesh up to twenty-five miles away.

When you gather together all the up-to-date evidence, you get a
very different picture than the one popularized by the specimen that
used to stand in the American Museum of Natural History. The
revised *T. rex* is a twelve-thousand-pound animal that can't grasp,
can't run, and can't see. Doesn't sound like much of a predator,
does it? But it needn't have been able to do any of these things if its
food weren't going anywhere, if, in other words, its "prey" were
already dead. What *T. rex* lacked in visual acuity was more than
compensated for by its exceptionally keen sense of smell, and its
massive legs would have served it well as it hiked long distances to
a waiting, ever-ripening meal. Besides, if the huge animal moved too
quickly it might stumble to the ground, crushing itself under its own

weight. *T. rex* might have been capable of running whenever it liked, but it would fall only once. Despite its size and ferocious appearance, then, the dinosaur many consider the king of its kind was an opportunistic scavenger, a connoisseur of carcasses, not an aggressive hunter like *Velociraptor*. If in *Jurassic Park* the Jeep had crashed and Ellie and Ian had died, the tyrannosaur might have sniffed out the site, but only after their bodies had been rotting long enough to broadcast a telltale odor. In any event, the dinosaur would not have pursued the Jeep as it sped away.

That *T. rex* behaves the way it does in *Jurassic Park* is no surprise, however. Animal aggression has always been misrepresented in popular accounts, and probably always will. Who would pay to see a shark movie, after all, in which the shark doesn't attack people? Where's the danger, the thrill in that? Late last summer I received a call from Jeff Goldblum, the actor who plays Ian in *Jurassic Park* and *The Lost World*. During his preparations for the sequel, a question had come up regarding the plausibility of the role. There are two groups of characters in the second story, those who hunt dinosaurs and those who gather information about them, and the dinosaurs frustrate the efforts of both. Goldblum, whose character is a gatherer, couldn't imagine how anyone might study living dinosaurs. "You can't get near them," he said. "They kill everything in sight." I explained that even the deadliest of creatures, those with the ability to inflict the greatest harm, do no more than what they are adapted to do. If a lion attacks a wildebeest, it does so because it's hungry, not because it's vicious by nature. If a grizzly bear chases a backcountry hiker, it does so to protect its territory, not because it's bloodthirsty. The same is true of dinosaurs. Even a hunter as effective as *Velociraptor* would have killed only as necessary for survival, and no more. In science, we call this parsimony, and we place a high value on it. We favor the simplest, most economical explanations when trying to understand the natural world, because that's what nature seems to favor. Fiction writers and filmmakers, by contrast, seek to entertain, and that is less likely to be achieved through parsimony than through imaginative elaboration—in this instance, by ascribing to animals behaviors that are in fact very difficult (though not impossible) to find outside the human sphere, such as deliberate cruelty, revenge, and random violence.

I've taken you on this detour because a dinosaur as popular as *T. rex* deserves to be known for what it actually was and because it shows how the science of paleontology actually works: by adding and subtracting bits of information, we gradually construct a coherent picture. When viewed against the backdrop of the entire movie, however, the dinosaur's predatory behavior is a minor distortion. In the popular arena, no effort to bring dinosaurs to life matches *Jurassic Park* and *The Lost World* for authenticity, to say nothing of sheer dramatic impact. Like Michael Crichton's books, on which they are based, the films draw heavily from the paleontological research of the 1970s and 1980s, which overturned much of what scientists had previously thought about dinosaurs. It is especially pleasing to see that the overall depiction of living dinosaurs is consistent with the newly revised view—that they were avian in the ways that mattered most, anatomically and metabolically, more akin to chickens and ostriches than geckos and monitors. If the films had been made as little as fifteen years earlier, many scenes, especially those dealing with eggs, embryos, and nests; parental care; and herd behavior, would not have been possible, for the simple reason that at the time these aspects of dinosaur existence were unknown, underappreciated, or misunderstood.

Anyone familiar with my research will recognize that the foregoing examples were derived in the main from discoveries we made in an arid expanse of rock and scrub brush called the Willow Creek anticline, located a few miles outside Choteau, Montana, all of which are described in my book *Digging Dinosaurs,* coauthored with James Gorman. The book covers fieldwork my crews and I conducted from the late 1970s through the summer of 1984, when we completed the excavation of Egg Mountain, an otherwise undistinguished hill located on the western slope of the Willow Creek site. There we had the good fortune to unearth three separate nesting grounds and a dozen egg clutches left behind by a small dinosaur called *Troödon**—the final stage of one of the most productive paleontological digs in history. The high point of that excavation, however, and the event that did more than any other to alter our understanding of dinosaurs, was the discovery of a nest containing the babies of another unfamiliar species, a duckbill, which provided the first direct evidence that adult dinosaurs reared their young, bringing them food, providing protection, and so on—in short, that they behaved very much like birds. In recognition of the creature's talent for parental care, my longtime friend and crew leader Bob Makela and I named it *Maiasaura,* or "good mother lizard."

Among the many good mothers that wander about the island enclave called Jurassic Park, none is more attentive to its progeny than the park's founder himself, the scientist who brought to life the *Velociraptor,* tyrannosaur, and other dinosaurs in the first place, and this is where the story takes leave of reality altogether. While *Jurassic Park* does a very good job of representing what we now know about the lives of dinosaurs, its underlying premise—that dinosaurs can live again—is fantasy in the first degree. The biggest challenge facing anyone wishing to clone a dinosaur from a set of dinosaur genes, or DNA, as the movie scientist purports to do, is not acquiring the DNA. We're in the process of doing that in our laboratory at the Museum of the Rockies in Bozeman, and we are attempting to do it the easy way, extracting dinosaur DNA, along with other organic molecules, from dinosaur bones. The movie scientist does it the hard way: he finds a seventy-million-year-old mosquito that happened to draw blood from a dinosaur, then happened to be trapped in sap, which thereafter happened to undergo the geological processes that turn sap into amber. I find fossil bones all the time, but I've never found a piece of amber that contains an ancient mosquito that in turn contains a piece of dinosaur. I doubt

---

*An abundance of skeletal material belonging to an even smaller, plant-eating dinosaur we named *Orodromeus* led us to believe that Egg Mountain was a nesting site for that dinosaur. Recent research, described in chapter 9, has persuaded us otherwise.

that I ever will, or that anyone ever will. And even if prehistoric blood were extracted from a prehistoric insect, it would be unlikely that we could determine whether the blood was sucked from a dinosaur or from a bird, since their genes probably are very similar.

Jurassic Park further compounds the make-believe by implying that DNA is sufficient for bringing extinct creatures back to life. Modern scientists clone by artifically dividing a fertilized animal cell into two fertilized daughter cells, then cultivating them separately, producing a pair of identical organisms. Despite its small size, an animal cell is an exceedingly complex network of organic chemicals, a sort of microscopic ecosystem, in which the magical strand of genetic code may be the most important component but is only one of a great number of necessary components, all of which interact and modify each other. At this juncture we don't know how the genes in the simplest of organisms function, much less how they operate within the machinery of the cell. And we surely are in no position to duplicate that exquisite machinery. We simply do not know how to induce DNA of any kind, dinosaurian or otherwise, to perform as it would normally perform when inside a cell.

That's not the only obstacle to cloning dinosaurs. Even if by some miracle we actually had a fertilized dinosaur cell, there would be nothing appropriate to grow it in. Mice and elephants can be cloned because female mice and female elephants are alive today. Fertilized animal cells can be placed in healthy animal wombs. And if duplicating a cell is so far beyond our capabilities that it belongs to the realm of science fiction, imagine the difficulty of duplicating a whole womb—or, more to the point, an egg, the place where dinosaur cells grow into full-fledged organisms. Rather, grew: between 230 and 65 million years ago, in the geological era known as the Mesozoic. Could dinosaurs be re-created and held captive in the twentieth century? Only in the minds of imaginative human beings. And that goes for paleontologists as well. I may know more than you about dinosaurs but I'm no more likely to see one. No, the feature that defines the study of dinosaurs, even lending to the endeavor a certain poignancy, is that the objects of our interest have vanished. They are long gone from this planet and they will never return.

Then why the fascination? Having collected the fossil remains of dinosaurs since I was a boy, devoted more than twenty years to speculating about their days on Earth, and worked as a consultant on both *Jurassic Park* and *The Lost World*, I've had ample opportunity to ponder their privileged place in the human imagination. Among extinct creatures dinosaurs have always held a special appeal, but now that we know they were less reptilian than avian they seem closer to us, slightly more approachable, and because of that all the more intriguing. Think of it: at least some of these huge, bizarre-looking creatures tended to their young. That is an incredible scene to contemplate. But beyond the interest aroused by the revised picture of dinosaurs there is the simple, overwhelming fact of their existence. In many respects dinosaurs resembled extraterrestrials, yet they walked the earth. They are the only large-scale aliens that we can be certain visited the planet, actually lived here, flourishing for an inconceivably long time. What better way to excite the mind than to invoke a partial picture of these somewhat familiar but fundamentally strange beings while inviting the imagination to fill in the gaps?

Dinosaurs are undeniably real yet deeply puzzling. This is what all of us fasten upon, whether we are novelists, filmmakers, full-time paleontologists, weekend collectors, or simply human beings doing what human beings never seem to get enough of—trying to figure out what's real and what's not. And we all start at the same place, with the bones—the fossil remains of dinosaur skeletons. Those, we know, are real. Where we go from there, how exactly we flesh out those skeletons, varies greatly, of course. But the aim is the same: to piece together pictures of dinosaurs, then to assemble enough pictures to be able to tell dinosaur stories. That's why I had no qualms about lending my name to the two epic dinosaur tales told by Michael Crichton and Steven Spielberg. It was clear that their depiction of dinosaur lives was going to be superior by far to anything that had been done before (the inaccuracies in earlier dinosaur movies are too numerous and, now, too irrelevant to address). But more than that, I applaud any sincere effort that might inspire people to give a little more attention to the many questions raised by dinosaurs. There can never be too much interest in a subject so large and so close to the mystery of existence itself—the very fact that there is life here at all and that everything that's alive today is so because everything else passed away. Indeed, as you probably know, our evolutionary ancestors, the early mammals, flourished only after, and perhaps because, the dinosaurs went extinct. So dinosaurs offer an unsurpassed window through which to view natural history.

Among all those who might step up to that window, trying to catch a glimpse of *Velociraptor, Tyrannosaurus, Maiasaura,* or any of their many cousins, there is one group of dinosaur enthusiasts that must adhere to strict rules, and that is the group to which I belong: paleontologists. Imagination surely plays a crucial role in my work—helping me fill in gaps, recognize patterns, and make guesses about where I should look for additional clues—but I differ from the nonscientist in that I can never stray far or long from the available evidence. In paleontology that means fossils—bones, for the most part, but also footprints, nests, any impression that might be preserved in or as rock. And it means rock itself, from microscopic mineral deposits to the large-scale structure and dynamics of the planet. The geological record contains the only traces of the environments where dinosaurs lived, and it's impossible to understand them without knowing something about their world.

There are two general approaches to paleontological research. You can find new ways to interpret fossils that have already been collected, an activity that sometimes is tremendously fruitful, or you can go out into the field and find more fossils. Like many other paleontologists, I do both, but chiefly the latter. And that's what this book is about—the fieldwork we've conducted since the summer of 1985, taking up where *Digging Dinosaurs* ended, as well as taking off in a new direction altogether. During the research that led to the discovery of nests, eggs, and babies, I played the role of animal behaviorist, specializing in the study of dinosaur families and other social groups, something that had never been done before. And I've continued in that role ever since. During the past twelve years we've uncovered thousands upon thousands of fossils that have helped us understand such behaviors as nesting, herding, migrating, foraging, and hunting. I'm biased, of course, but I believe that this research is among the most exciting in contemporary paleontology, and I'll devote a large part of the book to describing what we found, where and how we found it, and what it tells us about dinosaur behavior. By the late 1980s, however, it began to dawn on me—and this is where imagination comes in—that the tremendous amount of new evidence we had collected, especially the large bone beds we unearthed almost everywhere we looked in Montana, contained information about something far more important than behavior. I realized that the fossils we held in our hands were nothing less than clues to evolution itself—how certain dinosaurs came into being, developed and diversified, then went extinct.

I hope that by the end of the book you will appreciate and perhaps even share my enthusiasm for this turn of events. Meanwhile, as you read through the fieldwork section it might help to keep in mind that my ambition as a paleontologist is not to collect and catalog fossils, no matter how fascinating particular fossils may be. Sure, I like finding old bones, extraordinarily old bones, but my aim in searching for them is to decipher what they have to say about time, specifically, how life changes through time. I want to understand dinosaurs as living animals, to apprehend as much of their day-to-day existence as the evidence will allow. More than that, I want to understand how their lives unfolded from year to year, generation to generation, and era to era.

Loosely speaking, it can be said that the first paleontologists to take up the study of dinosaurs concentrated on developing portraits, trying to determine how certain individuals appeared and might have moved; more recent paleontologists like myself have been assembling dinosaur stories in an effort to show how dinosaurs behaved. This book goes a step farther, gathering together enough stories of behavior, especially group behavior, to be able to tell the sagas, or histories, of entire lineages of dinosaurs. So far, the only evolutionary issue that's been given serious consideration in the study of dinosaurs has been the manner of their disappearance, a much-overrated topic that I'll address later. But considering that dinosaurs flourished for more than 150 million years, successfully inhabiting the entire planet during that time, it would seem that the most compelling question is their longevity—their emergence, how they survived, the strategies that sustained them through a long and turbulent period in the planet's history. Until a few years ago, important clues to this evolutionary mystery lay buried in the immense dinosaur graveyards of Montana. This is the story of their discovery.

2
CAPTIVATING DINOSAURS

Six miles below the Canadian line, in the northeastern corner of the Blackfeet Indian Reservation, rests a prominent flat-topped hill called Landslide Butte. The Milk River cuts nearby, the water so sluggish and thick with sediment it looks like liquid chalk year-round. Steep gullies and ragged washes are carved into the scrub country between the hill and the border, everywhere exposing layers of shale, mudstone, and sandstone. It is in heavily eroded, largely barren, outcrop-rich regions like this, called badlands, that fossils are often found. And it is to this remote part of Montana that a pioneering paleontologist named Charles Gilmore came in search of dinosaurs almost eighty years ago. Over the course of several expeditions, he found a slew of fine specimens, including two new horned dinosaurs, which he named *Brachyceratops montanus* and *Styracosaurus ovatus*. Having seen the fossils that Gilmore brought back to the Smithsonian Institution, where he worked, I had long wanted to explore the Landslide Butte badlands. My chance came in the mid-1980s.

Map of Montana showing the approximate locations of our important discoveries. BTC = Blacktail Creek; LB = Landslide Butte; LS = Livingston Sauropod Site; MAL = Malta *Brachylophosaurus* Site; MRB = Milk River Badlands; T&J = Turner Ranch; TMR = Two Medicine River; TR = *Tyrannosaurus rex* Site; WCA = Willow Creek Anticline.

One sweltering day in the summer of 1986, I retraced his steps, relying on diary notes and photographs marked with arrows and *x*'s indicating the locations of his excavation sites. Paleontologists routinely keep field journals in which they record findings and observations, sketch specimens, draw maps, and otherwise document their work. I enjoy studying the journals of the early dinosaur explorers because they have an air of immediacy that cannot be duplicated in more leisurely, retrospective accounts. And sometimes they reveal aspects of the scientist's thought processes and research methods that otherwise might remain hidden. In his notes, as well as in later publications, Gilmore described a layer of mudstone that is exposed throughout the Landslide Butte badlands. The layer, he wrote, is unusual in that it is littered with the fossilized shell fragments of a freshwater clam.

Gilmore's clam layer interested me because he had used it as his datum level, his point of reference in the local geologic column.

Since sediments like sand and mud are deposited in sequence, newer layers atop older ones, a vertical column of sedimentary rock serves not only as a record of what was deposited but also when it was deposited in relationship to everything before and after it. If a certain layer is distinctive in some way, that is, easily identifiable, and extends over a large enough area, it can be used to gauge the age of surrounding sedimentary rock and any fossils they might enclose. By measuring the distance between the benchmark, or datum, and layers where dinosaurs are found, you can determine when the dinosaurs lived, sometimes with surprising precision.

What surprised me that summer day eleven years ago, however, and what would probably flabbergast Gilmore were he alive today, was what I stumbled upon when I examined his datum layer at close range. I was sitting on a hill, comparing one of his photographs with the rock in front of me when I saw that the shell fragments were not those of ancient freshwater clams but instead of eggs, dinosaur eggs, objects that the uninitiated often overlook or misinterpret but which I had had plenty of practice identifying at Egg Mountain, on the Willow Creek anticline. Incredibly, Charles Gilmore had found an extensive deposit of dinosaur egg remains in Montana in 1916, among the first in the world to do so.* But he didn't know it. He didn't recognize the fragments for what they really are. Consequently, official credit for the discovery of dinosaur eggs went to the adventurer Roy Chapman Andrews, during his American Museum of Natural History expedition to the Gobi Desert in Mongolia in 1922.

I haven't related this incident to show how clever I am at the expense of a fellow paleontologist who can no longer defend himself. Gilmore's misinterpretation doesn't mean that he was an inept scientist. It just means that he practiced a science in which misinterpretation, along with oversight and plain bad luck, are occupational hazards, and as much so today as back then. I shudder to think how many times I may have stared at a monumental find and failed to see it, or stood on top of something I'd been pursuing for weeks,

---

*Eggshell fragments had been found earlier in France, but also without being recognized as such.

Photograph of Charles Gilmore and George Pearce working in a bone bed of juvenile hypacrosaurs (Gilmore referred to them as procheneosaurs, believing them to be small adults rather than juveniles) near Landslide Butte in 1935. (*George F. Sternberg, reproduced courtesy of the Smithsonian Institution*)

then walked away without realizing how close I actually was. The day I reenacted Gilmore's journey through the badlands of Landslide Butte I had a huge advantage—in my mind I carried what I like to call a "search image," in this instance a clear picture of dinosaur eggs, derived from years of experience collecting them, studying them, analyzing the rock in which they are likely to be found. What's more, in 1986 that's precisely what I wanted to find; I was driven in that direction. But had I arrived at Landslide Butte when he did, having never heard of anyone finding an egg, much less actually seeing one myself, I probably would have come to the wrong conclusion about those fragments, too. In paleontology the first person to make a discovery invariably finds himself in a less advantageous position than those who follow in his footsteps, because he always has less information than they will. Ideas are continuously overturned in paleontology not because paleontologists are sloppy or incompetent but because they continuously unearth new fossils, new rock, new data, and that leads to revision.

This feature distinguishes paleontology, whose subjects are living, evolving organisms that are no longer alive, from most other scientific disciplines, but especially those, like physics and chemistry, that traffic in the immutable laws of nature, which are everywhere and always the same. There is no equivalent in paleontology to the law of gravity, no equations that apply to the behavior of one kind of dinosaur under one set of circumstances, still less to all kinds under all circumstances, no mathematical procedures for predicting exactly where or how fossils will be deposited. Moreover, unlike botany or zoology, which also concern living things, paleontology is a historical science, a science based on circumstantial evidence, after the fact. We can never reach hard-and-fast conclusions in our study of ancient plants and animals, points beyond which no further debate or research would be necessary. These days it's easy to go to school for a good many years, sometimes even through college, without ever hearing that some sciences are historical or by nature inconclusive. But in fact paleontology is closer in spirit to the traditional definition of science—a method rather than a set of principles, a form of systematic doubt, a way of testing ideas.

If you've read many popular accounts of the scientific enterprise as it's practiced today you probably have gotten tired of hearing this or that discipline described in terms of detective work, with sleuths scouring the world over for clues to their particular questions. But if there's any branch that deserves the comparison it's paleontology. It is, after all, the only field that actually deals with dead bodies. The one great advantage I have over a real detective is that the bodies that interest me are long dead, so long that the flesh has dissolved and the bones have been replaced by stone. In short, their remains do not smell. The great disadvantage is that the bodies have been dead so long that nothing quite like them exists today. Everything about their lives, and deaths, must be inferred indirectly from a relatively small amount of extremely old evidence. I'll never see any of it with my own eyes. It's more than a matter of trying to comprehend events that occurred in the past. The world in which those events took place has vanished as well, much of it lost forever, never to be recovered in any form, not even as an idea.

I happen to believe that we may one day discover a universal theory of evolution—something resembling the law of gravity—but no doubt that day is a long, long way off. Until then, and probably afterward, dinosaur paleontologists will be obliged to proceed as, yes, detectives, collecting clues wherever and whenever the opportunity arises, concocting one idea after another, constantly revising, in a ceaseless and open-ended process of approaching—and I must emphasize the word *approaching*—the truth. Picture the seemingly disorganized television character Columbo turning around for the umpteenth time and saying, "Oh, I forgot, just one more thing." That's the paleontologist, only in perpetuity, never without an unanswered question, a further qualification, another way of looking at things, forever getting closer but never quite arriving.

And therein, ironically, lies his pleasure, for in paleontology there is always something more to be learned. We may never reach conclusions, but by the same token we will never cease making new discoveries either. Paleontology may be inexact, tentative, even contradictory, but it is perennially dramatic. There's always an atmosphere of anticipation surrounding what we do. Since the turn of the century, by contrast, physicists have regularly pronounced their discipline dead, or nearly so. All of the fundamental laws of nature have been figured out, the argument goes, and there's nothing left to do but sweep up and turn out the lights. It is inconceivable that a paleontologist would make such a statement, that the field would become so exhausted of opportunity that its practitioners might begin contemplating retirement. As a paleontologist I have to live with a large measure of uncertainty, but I also live in the conviction that a marvelous find awaits me just over the next hill, on the next slide I examine under my microscope, in data reconsidered from a new perspective.

As befits a historical science that is concerned more with dynamic processes like organic development and evolution than with static principles like the law of gravity, paleontology is best appreciated in terms of its own development. Toward the end of the next chapter I'll tell you more about Gilmore and some of the other twentieth-century explorers who pioneered dinosaur paleontology in Montana, one of the most fossil-rich regions in the world, but right now I'll recount a few episodes from the earliest days of the field. You can get a pretty good sense of what physics and chemistry are all about by studying them as bodies of knowledge, concentrating on what is known to be the case today. Getting acquainted with the history of each field will enhance your appreciation, certainly, but it isn't necessary. The spirit of paleontology, on the other hand, is almost impossible to grasp without some familiarity with its development. The day I relived Gilmore's exploration of the Landslide Butte badlands I was doing what every paleontologist does every day of his or her life. If, as it is often said, contemporary physicists stand on the shoulders of those who have gone before them, then contemporary paleontologists walk in the footsteps of those who have gone before, literally and figuratively. We constantly retrace well-worn tracks and reopen old excavations. More important, every notion we take up is but a variation on a single theme: the idea that organisms do in fact change, develop, evolve, that nature has a history.

Back when dinosaur remains were originally discovered, the idea of natural history would have been infinitely more alien than the fossils themselves. It's not known, of course, when or where exactly a human being first contemplated the skull of a dinosaur, but the earliest encounter on record can be inferred from a legend the Greeks borrowed from Central Asia in the seventh century B.C. According to Adrienne Mayor, a classical folklorist who only a few years ago pieced together this amazing story from a wide range of sources, pre-Christian oral tales from Greece and Rome describe the griffin, a half-bird, half-mammal that guarded caches of gold in the Altai Mountains along what is now the border between Mongolia and China. Griffins walked on all fours and had wings that originated in the shoulder region, a horn rising from the top of the head,

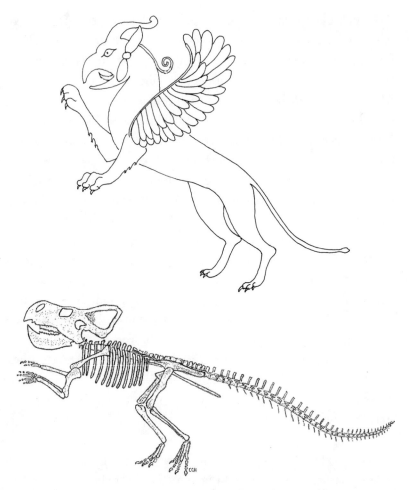

The griffin, considered a mythical creature by western culture, was most likely based on the skeletons of the dinosaur *Protoceratops,* common in the rocks of the Gobi Desert, near the Altai Mountains, where griffins are alleged to have existed.

and a prominent beak. Greek art depicts both adults and juveniles, the adults even protecting their young.

As Mayor discovered after consulting with paleontologists, including my friend Philip Currie of the Royal Tyrrell Museum of Paleontology in Alberta, Canada, it just so happens that immediately south of the Altai Mountains, a highly productive gold mining region (*Altai* means gold in the local dialect), lies the Gobi Desert, where a particular section of sedimentary rock known as the Nemget Formation is exposed. The formation is chock-full of fossils. It is where Roy Chapman Andrews and his crew found the world's first clutch of dinosaur eggs in 1922. It is also where they found the first skeleton of a relatively small horned dinosaur called *Protoceratops,* followed by a great many more. During the course of two summers Chapman and his crew excavated upwards of one hundred of them. *Protoceratops* is by far the most common dinosaur in the Nemget Formation. In a relentlessly arid, wind-ridden environment where fossils tend to be unusually well preserved, *Protoceratops* skeletons are also unusually easy to find. The bones are white, whereas the desert rock in which they are lodged is bright red. And it has been this way for countless millennia, the bones of *Protoceratops* weathering out of the sandstone cliffs for anyone passing by to see, like the nomadic peoples who inhabited the area twenty-seven hundred years ago.

Here is the intriguing part: a *Protoceratops* skeleton could easily be mistaken for the remains of a griffin. For one thing, the dinosaur has a very prominent, birdlike beak. For another, an equally prominent bar, or frill, extends from the back of the skull to the neck shield. But the section near the shield is so thin that more often than not it breaks off and disappears long before anyone finds the skull, leaving behind a structure resembling a horn, very much like the horn depicted in Greek and Roman pictures of the griffin. The flared edges of the frill might well have been the inspiration for the griffin's long ears. Mayor speculates that as the legend of the creature was passed along, wings were added to complement its avian features. This could be true, but *Protoceratops* has an elongated shoulder blade located in exactly the same place as the griffin's wing, which might not seem significant until you consider that the wings of mythical creatures typically are located in areas where no bones are present to anchor them. The next time you see an angel, especially one blessed with big wings, ask yourself which anatomical structure supports those heavy limbs.

I think Adrienne Mayor has made a convincing case that the legendary griffin, which remained part of the Greco-Roman artistic and literary traditions until the third century A.D., was inspired by the *Protoceratops* skeletons of the Gobi Desert. Back then most of the world was terra incognita and, by today's standards, cultures had very little contact with each other. The realm of the might-be-real was almost without limit. If you came across a *Protoceratops* skeleton, or any other unusual skeleton, there was every reason to believe that similar animals still existed, if not in the immediate area, then somewhere else. What's more, there was nothing to suggest that the group to which an unfamiliar animal belonged might have died out.

As far as anyone can tell, the concept of extinction was completely unknown at the time, and remained so until explorers had charted the continents in some detail and the peoples of the world became acquainted through commerce and cultural exchange. From the dawn of humanity until that point, the standard belief was that all the plants and animals on Earth had been created at the same time and that all of them had continued to exist in the same way from that day onward. Life, in this view, is unchanging. Even events as catastrophic as the biblical Flood are incapable of upsetting the natural order. Most of the world may have been inundated but, with a little divine prodding, ship-building Noah saved the animals from destruction. If nothing else, the persistence of this notion demonstrates that ideas need not agree with the available evidence to have enormous influence.

And persist it did, into the 1600s, when a few European naturalists openly speculated that since living animals could not be found to match certain skeletons, the animals must have perished sometime in the past. But it wasn't until the latter half of the following century that the idea of extinction really took hold. For decades laborers in the chalk quarries near Maastricht, Holland, had been finding troves of marine fossils, from spiral-shelled ammonites to stony urchins, which wouldn't surprise anyone today, since chalk is a type of limestone comprised almost completely of seashells. At a place called St. Peter's Mountain the pale rock was so plentiful that quarries had been expanded into an elaborate complex of tunnels and subterranean galleries, where collectors often searched for fossils by torchlight. And there, in 1770, deep within the mountain's fossil-laden interior, miners discovered a gigantic set of jaws unlike anything seen before. At the time, opinion varied widely on the origin of the stupendous and much-prized fossil, some believing that the jaws belonged to a prehistoric whale, others a marine lizard, and still others an ancient crocodile.

Twenty-five years later, after a bitter legal battle over rightful ownership and a perilous, roundabout journey through the French Revolution and subsequent Reign of Terror, the jaws arrived in Paris, where they were examined by the brilliant young anatomist Baron Georges Cuvier, who was well positioned to render an enlightened judgment about the beast that had come to be known as *Mosasaurus*. Having taken it upon himself to answer once and for all the controversial questions raised by the existence of fossils, Cuvier was just then completing a comprehensive survey of elephant and elephantlike remains. Comparing the skeletons of the American mastodon and the Siberian mammoth with those of Indian and African elephants, he demonstrated that the animals differed anatomically in small but significant ways, which meant that they probably lived differently as well. Since no one had ever seen a mastodon or mammoth alive, he reasoned, the creatures must have gone extinct. Cuvier also demonstrated that the further back in time one travels, the greater the gap there is between extinct organisms and those that exist today. The mammoth, for instance, was found in relatively young rock and in the main resembles the contemporary elephant, whereas the mosasaur came from much older sediments and bears an overall anatomy that differs strikingly from its closest living relative, the monitor lizard. The mosasaur's long, toothy jaws were well-suited for eating fish, it possessed fins instead of feet, and its tail was designed for propulsion through water. The creature surely enjoyed a different existence than its terrestrial cousin—in the sea.

To account for the seemingly sudden disappearance of one species and the equally sudden appearance of its replacement, Cuvier proposed natural disasters of biblical proportion. This explanation placated the religious community, already deeply offended by the notion of extinction and what it implied about Creation, but it became increasingly difficult to defend as the number of acknowledged extinct animals—and, thus, precipitating global catastrophes—increased. Of course, Cuvier was mistaken, as Charles Darwin made clear when in 1859 he published *On the Origin of Species*, firmly establishing the idea of evolutionary succession by natural selection. But by interpreting, correctly, the first fossil leviathan as a giant marine lizard, Cuvier inadvertently lent his name to another misconception, one that haunted paleontology for more than 150 years. When early in the nineteenth century two newly discovered terrestrial leviathans were given official names, *Megalosaurus* and *Iguanodon*, it was assumed that they, too, were saurians—lizards. Then, in 1841, the British paleontologist Richard Owen successfully argued that the lower vertebrae of these two creatures, and of a third, called *Hylaeosaurus,* were different enough from those of other giant lizards that the three animals should be considered a separate suborder of reptiles, which he called *Dinosauria.* Not only were the new creatures lizards, they were terrible lizards.

The name stuck, as we all know, and so, unfortunately, did the many unexamined assumptions associated with it. From that point onward paleontologists expected dinosaurs to be reptilian, that is, cold-blooded, sluggish, dim-witted. That was their search image, so that is exactly what they found. Anything that might have contradicted the image was overlooked, misinterpreted, or dismissed as unimportant. Nor did paleontologists show much interest in such contemporary animals as the crocodilians, which are the reptiles to which the dinosaurs are most closely related, and that too helped perpetuate the "terrible lizard" image.

My purpose in providing this extremely abbreviated account of the early days of dinosaur paleontology is to make two observations whose relevance will continue throughout the book. First, the idea of evolution—living things change through time—is a recent one, which may help explain why it's still widely misunderstood and misrepresented. Second, we tend to see the unfamiliar in terms of the familiar, in this instance, apprehending dinosaurs by overemphasizing their resemblance to reptiles. I hasten to add that this is a mistake that's difficult to avoid. The fact is, dinosaurs are related to living animals. That's what we mean when we say that life has evolved. Some similarities, then, are to be expected. The challenge for the first generations of paleontologists was to acknowledge the validity of that expectation without permitting it to blind them to very real differences.

Perhaps if the animals had been given another, more biologically neutral name from the start, one that placed less stress on their superficial reptilian features, they would have been seen for what they really were much earlier. As it happened, though, only in the late 1960s did it become widely accepted that the dinosaurs' closest living relatives may not be the reptiles after all. Among the many scientists who contributed to this revolution, the most influential was John Ostrom of Yale University, who had uncovered a group of carnivorous dinosaurs in south-central Montana in 1964. *Deinonychus*, as Ostrom called the animal, was about eight feet long, and had well-developed scaffoldings of tendons supporting both its backbone and tail, suggesting that it stood upright, with its tail erect and off the ground. Judging from this and other crucial evidence, including its long, powerful legs and distinctive toes, one of which had evolved into a deadly claw, Ostrom concluded that *Deinonychus* "must have been a fleet-footed, highly predaceous, extremely agile and very active animal, sensitive to many stimuli and quick in its responses. These in turn indicate an unusual level of activity for a reptile and suggest an unusually high metabolic rate." (If *Deinonychus* calls to mind the *Velociraptors* of *Jurassic Park* and *The Lost World*, it is no accident; they're closely related.)

Unusual level of activity. Unusually high metabolic rate. The door had been opened, and opened wide, by a paleontologist with an unimpeachable reputation, and thereafter one revisionist after another walked through, bringing news about the dinosaurs. Some of them, it seems, were warm-blooded. Others were highly social. One group of duckbills, also found in Montana, actually built nests, lived in colonies, and cared for their young. Ostrom himself showed that in several important ways *Deinonychus* was anatomically similar to *Archaeopteryx,* the first bird; that indeed birds in general descended from dinosaurs, which was the culmination of one of the most dramatic turnarounds in modern scientific thought. It now seems that the winged griffin of the Gobi Desert, that fanciful hybrid of avian and mammalian features, struck closer to the truth than anyone could have guessed.

I'll return to the subject of evolutionary relationships, including those between dinosaurs and birds, after I've described the field-work I conducted during the past ten years and discussed how our ideas about evolutionary relationships affect fieldwork's direction and outcome. The persistent misinterpretation of dinosaurs overall, like Gilmore's misidentification of dinosaur eggs at Landslide Butte, calls attention to the major pitfalls of trying, on the basis of incomplete and indirect information, to reconstruct lives that ended tens of millions of years ago—reading either too much or too little into the evidence at hand. One way to guard against this pitfall is to practice something scientists call the Null Hypothesis, which means simply to actively seek evidence that might contradict what we would like most to believe. In criminal detection this approach is known as the process of elimination, by which an investigator pares away possible explanations for a crime. Because it runs counter to the normal human tendency to look for confirmation, the Null Hypothesis helps keep scientists honest. The broader implication is that science advances as much through the invalidation of ideas as through their validation. Though it was slow in coming and largely accidental, there is no better example of progress through invalidation than the replacement of the reptilian interpretation of dinosaurs with an avian one. Paleontology is not a field for people who want to be right, it is a field for people who want to know what's real. Now that we have learned how easy it is to go wrong, and once wrong, continue to go that way for an embarrasssingly long time, it's all the more important that we not wait for time to overturn our most cherished notions but instead look for contradictory evidence ourselves. Sound ideas will survive such scrutiny.

These thoughts were going through my head back in the 1980s, when we first made public our ideas about nesting and parental behavior. A number of paleontologists immediately challenged the underlying research. In discussions with Bob Makela about the direction of fieldwork following the excavation of Egg Mountain, I told him that the best way to answer the critics would be to continue looking for baby dinosaurs, but with an eye toward uncovering evidence that might invalidate the stories we had been telling about dinosaur behavior. I had convinced myself that adult maiasaurs took care of their young, but it was exactly that conviction that might prejudice my interpretation of additional information, especially if the information were contradictory. So I proposed to Bob that we deliberately try to disprove the very ideas we had been advocating. In the opening chapter I introduced these ideas (for those familiar with *Digging Dinosaurs,* reintroduced them), but I'm now going to recount, in a little more detail, the evidence on which they were based, so you can better appreciate why they aroused suspicion among some of my colleagues as well as what was at stake when I expanded my search for eggs and babies to other parts of Montana.

Of all the discoveries we made during seven seasons of excavation at the Willow Creek anticline, the most dramatic occurred at the outset, in 1978, long before we had any inkling of the magnitude of the fossil deposit on which we stood. Bob and I excavated a group of fifteen baby *Maiasaura* skeletons, each very nearly the same size, in a bowl-shaped depression that we soon realized was a nest. The next year we uncovered more nests, all located in the same layer of sedimentary rock, suggesting that they were constructed at the same time. Maiasaurs, it appeared, gathered in colonies to lay their eggs, very much like modern birds. Nearby we ran across addi-

Maiasaura peeblesorum sitting next to its nest full of babies.

tional evidence for social behavior—a bone bed containing the remains of at least ten thousand adult *Maiasaura* skeletons, all of the animals apparently killed at the same time by a heavy rain of volcanic ash. We had discovered, in other words, a herd of dinosaurs, herbivores, or plant eaters, that very probably migrated with the seasons, foraging on flowering plants, which had only recently evolved and were then dispersing across the continents.

There was more. The *Troödon* nesting sites on Egg Mountain meant that that relative of *Velociraptor* also nested in colonies. At Egg Island, site of another rookery, we found nineteen fossilized *Troödon* embryos, as well as the remains of *Orodromeus*. Most interesting, from the standpoint of parental care, were the differences between the bone joints of the embryonic *Troödon* and those of the baby *Maiasaura*. The *Troödon* joints were very well developed, indicating that they were precocial, that is, capable of getting around on their own by the time they were born. In all likelihood, as soon as *Troödon* hatched from its egg it clambered out of the nest, which explains why we found only embryos in the colonies. But the baby *Maiasaura* joints were poorly formed, immature, strongly resembling those of altricial birds, which remain in their nests and are fed and protected by their parents until their limbs grow strong enough to support them. To my mind, this was the strongest evidence of all. Baby maiasaurs simply could not have survived without lots of help from adults—without parenting.

Some of these finds—babies in a nest, herds of adults, the clutch of embryos—were the first of their kind in the world. Morever, in the years immediately following our excavation of the Willow Creek anticline, no one else found comparable specimens anywhere. That made some paleontologists skeptical, and especially those who, despite the revised view of dinosaurs, have not been convinced by the avian interpretation. Regardless of the views of my critics, however, I had to admit that the evidence for parental care was pretty slim, too compelling to argue away, in my view, but slim all the same. Stripped to its essentials, all we had was two sets of baby dinosaurs in nests and some differences in joint anatomy. I wanted more, to lay to rest not only the criticisms of others but my own doubts. And I wanted to apply the Null Hypothesis, the process of elimination, locating as many eggs and babies as possible in an attempt to show that dinosaurs did not take care of their young. The Willow Creek anticline had become so popular that we couldn't have mounted another serious dig there without hiring an army of security guards and public relations agents. Besides, we had already crawled over the area, literally crawled, alongside the ants, inch by back-splitting, knee-grinding inch, an entire square mile of hardpan. It was time to look elsewhere.

And that brings me back to Gilmore's datum. He had used the shell fragment layer, you will remember, as a chronological benchmark against which he measured the age of the fossils he uncovered in the badlands of Landslide Butte. All field explorers need such geological signposts; they need to know "when" they are no less than where they are. Our extensive excavation of the Willow Creek anticline west of Choteau provided me with a less specific but nonetheless clearly recognizable geological signpost—the suite of sedimentary rocks known in Montana as the Two Medicine Formation, in particular, the layers in which we found *Maiasaura, Troödon,* and *Orodromeus.* If we were going to find additional eggs and babies, we would most likely do so in similar geological matrices, which is to say, comparable levels of the same formation. Like most paleontological investigations, then, our expanded search for eggs and babies focused first on locating certain types of rock.

As it happens, Montana is one of the best places in the world to find dinosaur-bearing sediments. In fact, about one-quarter of the state's land surface contains rock from the Mesozoic era, and of that rock, a sizable portion is in the Two Medicine Formation. That would be our signpost. More important still, the formation would serve as our window on the environment *Maiasaura, Troödon, Orodromeus,* and their contemporaries lived in. The rock held clues to the physical forces that acted on the dinosaurs' lives, to their behavior, to their fate. It is impossible to interpret fossil remains without being able to read the rock in which they are found.

That's why I'm going to devote one more chapter to laying the foundation for the fieldwork section, a chapter on geological literacy, emphasizing in particular how to read the sedimentary rocks of Montana. If you wish to understand dinosaurs as something more than strange-looking skeletons you must see them in their native context, in their own world, as characters within stories. The stories are written in the rock that entombs them.

3

DINOSAURS LOST, DINOSAURS FOUND

Time, like beauty, exists in the beholder's eye. When the historian talks about bygone days he means the past twelve thousand or so years, the era of human civilization. To the archaeologist, the immediate past extends back a million and a half years, when hominids, or man-apes, first appeared. And when a paleontologist says that such-and-such happened just yesterday, he might be referring to last Tuesday or a Tuesday hundreds of millions of years ago. I'm exaggerating, of course, but only a little, and to make an important point: that one's sense of time—how old things are and how fast events take place—can vary greatly, depending on all sorts of factors. An everyday example is the profound change most of us undergo in our perception of time as we move further away from birth and closer to death. To the average teenager, days, weeks, and months unfold at a leisurely pace and years come and go so slowly that it seems like one will live forever. But that same person, once he is well into his middle years, is likely to experience days ticking by like minutes and years turning over like months. At sixty-five, life appears alarmingly shorter and time a great deal faster than they did at fifteen.

Why exactly this is so is a mystery philosophers have been trying without success to figure out ever since man became conscious of the passage of time, capable of remembering the past and anticipating the future. I won't try to improve on their efforts here. Besides, for the paleontologist it's enough to know that however long or short his own life may seem, it's but an instant, or less, when compared with the entire duration of life on Earth—about two billion years. To get a clearer sense of the comparison, imagine that that two-billion-year period is equivalent to one year, in the same way that a full-size *Ultrasaurus,* a sauropod that stood about six stories tall and weighed as much as 150 tons, is equivalent to a one-foot-high scale model. If that were the case, each month of the model year would represent about 165 million years; each day, about 5.5 million years; and each minute, about 3,800 years. My life so far—I just turned fifty-one—represents ⅕ of a minute. Even if I'm lucky and reach old age, my entire existence, from birth to death, will last somewhere in the neighborhood of one second. Think of that: one second in the year that life has existed on Earth. Come and gone in the blink of an eye.

Now try thinking of the scale model of time this way: During the year that stands for two billion years, dinosaurs emerged on about November 18 and went extinct almost exactly a month later, on December 18. Consider, by contrast, human beings. As I said earlier, hominids have been around for at least a million and a half years. That's equivalent to about six and a half hours on the scale year. To the best of our knowledge, the group of hominids to which we belong, *Homo sapiens sapiens,* or modern man, appeared about forty thousand years ago, which is equivalent to about ten and a half minutes. In other words, human beings made their first appearance on the evolutionary stage ten minutes before midnight on the last day of the last month of the year. Whenever we get carried away with notions of our special status among all of the creatures that have made this planet their home, it would be well to remind ourselves of this fact—that we are newcomers, and that we have a long way to go before we can say that we are one of life's success stories.

What does this have to do with dinosaurs? For one thing, it helps to show that when measured against any human-based scale—from an individual life span to the entire duration of the species—dinosaurs lived for an extraordinarily long time and, what's more, they did so an extraordinarily long time ago. For another, and I can't emphasize this enough, it helps one appreciate the breadth and potential fertility of evolution. Evolution is nothing more, and nothing less, than change through time, but to grasp the extent of possible change, as well as the mechanisms responsible, you have to cultivate a much-expanded sense of history. Only within a historical context, against the backdrop of their life stories and generational sagas, do dinosaurs reveal their full significance. To make what I'm saying more concrete, picture what paleontological fieldwork would be like in the absence of historical awareness, if it were merely a matter of finding and sorting fossil bones.

Last summer, for example, I conducted a fascinating excavation outside Malta, in northeastern Montana. Nate Murphy, an amateur collector from the area, had been exploring the eroded benches along one side of a broad, grassy drainage when he spied a dinosaur tailbone protruding from a wall of exposed sandstone. Since the fossil was located on public property, administered by the Bureau of Land Management, he couldn't excavate it without a BLM permit, and he didn't qualify for that because he's not a professional paleontologist. So he came to me and asked if he might work the site under my permit. I consented, but that made me ultimately responsible for the dig, so I drove to Malta to take a look. By that time Nate had uncovered the entire tail, and it was immediately clear to me that he had probably stumbled upon a museum-quality specimen—complete, intact, undisturbed—of an adult duck-billed dinosaur, which does not happen often. Since the sandstone was soft and crumbly, my crew and I were able to expose the entire

Excavation of the Malta *Brachylophosaurus*. The nearly complete skeleton is lying on its right side. The tail is to the left, the head to the right. (*Bruce Selyem, reproduced courtesy of the Museum of the Rockies.*)

skeleton in less than a week. And what a skeleton it was—a beautifully preserved, twenty-foot-long *Brachylophosaurus* that looked as if it had lain down on its right side and gone to sleep, never to rise again. Almost every bone was in place, even the fingers. The rib cage, which is crushed and flattened during the burial of most specimens, was inflated, bowed, just as if the animal still contained its internal organs. All of the tendons that kept its tail erect were present and in place.

Yes, a beautiful specimen. But what did it tell us? What was the story behind this particular brachylophosaur? By studying its anatomy, the shape of the bones and how they are joined together, we could get a pretty good idea of how the dinosaur moved. And it is true that by examining the makeup and internal structure of the bones under a microscope we could probably determine how fast it had grown prior to death. But where did the brachylophosaur come from? What was it doing here? What was its place in the larger scheme of things? Regarding these questions the skeleton was stubbornly silent. If we had to rely on the fossils alone, we would find ourselves in much the same position as the nomadic people who had discovered protoceratopsian skeletons in the desert of Outer Mongolia thousands of years ago. Like them, we could only guess at the origin of the unfamiliar animal, its relationship to other creatures, its current whereabouts. And like them, we'd probably invent an interesting tale for which there is little or no evidence—that the reason we don't see brachylophosaurs running around Malta today, for instance, is that they live deep underground.

If that scenario strikes you as something Jules Verne might have dreamed up, it's only because during the past two centuries paleontologists, historical geologists, and evolutionary biologists have been shedding light on the least understood dimension of life on Earth—time. Because of their efforts we now know that the sandstone surrounding the brachylophosaur is a physical record of the passage of time. To be precise, the outcrop from which we removed the duck-billed dinosaur skeleton belongs to a section of sedimentary rock called the Judith River Formation, and judging from the age of the outcrop we know further that the brachylophosaur died about 76 million years ago. And that's not all. The Judith River Formation consists of terrestrial sediments deposited while an inland sea shrank, its westernmost shore retreating steadily to the east, away from the Rocky Mountains, while it expanded again. In other words, the brachylophosaur inhabited the plains during a time when the plains had grown significantly wider, opening up new habitat for dinosaurs and other organisms, and thus making possible a great diversification and dissemination of life along the Rocky Mountain Front.

This isn't all I see when I contemplate the *Brachylophosaurus* skeleton in context—that is, against a historical backdrop—but it is enough to demonstrate why we sometimes spend as much time exploring the rock in which fossils are found as we do studying the fossils themselves. It should also convey some notion of the total search image paleontologists employ in the field. When we hunt for dinosaur bones we picture more than the rock in which we're likely to find them; we picture the world the dinosaurs inhabited when they were alive. In the largest possible sense the world we have in mind—that anyone interested in dinosaurs should have in mind—is the world of the Mesozoic era, from 230 million years ago to 65 million years ago, a time when enormous changes occurred on the surface of the planet, affecting all plants and animals, including where they lived, how they existed, and the overall course of evolution.

From the standpoint of the planet as a whole, the most significant geological event of the Mesozoic era was the breakup of Pangaea, the landmass into which all of the major continents had merged—a single colossal island in a single global ocean. Although exact dates are impossible to come by, geologists now generally believe that Pangaea remained intact until at least 220 million years ago, well into the Triassic, the earliest of the three periods that make up the Mesozoic era. This is important because it means that the first dinosaurs, not to mention the first mammals, appeared when, in principle, at least, animals could migrate from one "continent" to another without having to skirt large bodies of water. On this basis it would seem reasonable to assume that dinosaur fossils should be found throughout the world today, and in fact they are. When, at the outset of the Jurassic period, 195 million years ago, Pangaea began disintegrating, the early dinosaurs separated as well, some inhabiting Laurasia, the northern complex of continents, others inhabiting Gondwana, the southern complex.

The Jurassic period lasted about sixty million years. What was left of Pangaea continued to drift apart, with Gondwana starting to split into South America, Africa, India, and Australia-Antarctica, and a division beginning to show in Laurasia, the first sign of what would become Europe and North America-Greenland. Along that division the Atlantic Ocean eventually formed. During the Jurassic the continents were relatively low-lying. Worldwide the climate was warm and humid. Large expanses of Europe and, later, North America soon became submerged beneath shallow inland seas.

Gymnosperms, or nonflowering plants, were plentiful everywhere, with conifer forests occupying the uplands and ferns, giant horsetails, and large palmlike plants called cycads growing in the wetter, more tropical lowlands. Also well established by this time were the two great orders of dinosaurs: the Saurischia, or lizard-hipped dinosaurs, and the Ornithischia, or bird-hipped dinosaurs. Prominent saurischians included the sauropods, the largest and tallest land animals ever to have lived, and the theropods, the group of bipedal, flesh-eating dinosaurs to which *Tyrannosaurus* belongs. The plate-backed stegosaur and such primitive ornithopods as *Camptosaurus* were among the ornithischians that flourished during the Jurassic period.

Although dinosaurs became the dominant land animals during the Jurassic, and for that reason the period is considered the zenith of dinosaur evolution (a perception reinforced by Crichton's books and Spielberg's movies), the Cretaceous period, beginning 136 million years ago and ending 65 million years ago, has proved to be a great deal more interesting—for my purposes, at any rate. Three crucial features define the last period of the Mesozoic era. First, the continents continued to drift away from one another, creating increasingly larger oceans between them, while vast inland seas expanded and contracted in slow, rhythmic pulses in Europe and North America (where the sea was contracting when our *Brachylophosaurus* was alive). Second, extensive mountain ranges rose along the western coasts of both Americas, accompanied by violent, often long-lasting volcanic eruptions. Of particular interest are the Rockies of North America, which profoundly altered the climate of that continent by preventing rain from reaching its interior regions. Third, flowering plants, the angiosperms, came into their own, dispersing across the continents and diversifying into all manner of environmental niches. For the dinosaurs that ate plants (and most of them did), this represented not only a new and prodigious source of food, but a source of food that perpetually renewed itself through the annual replacement of leaves.

That is how the Mesozoic era looks when viewed all at once,

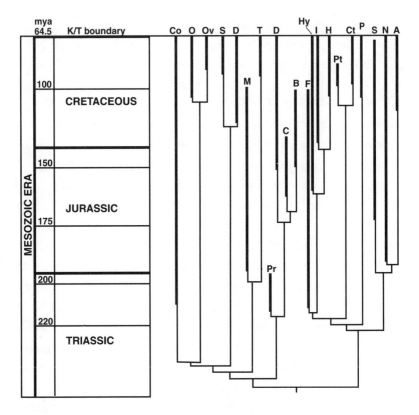

Diagram showing the origination and duration of the best-known dinosaur groups. Co = "Coelurosaurs"; O = ornithomimosaurs; Ov = oviraptors; S = saurnithoides; D = dromaeosaurids; M = megalosaurids; T = tyrannosaurids; Pr = prosauropods; D = diplodocids; C = Camarasaurids; B = Brachiosaurids; F = fabrosaurids; Hy = hadrosaurids; Pt = ptsittacosaurids; Ct = Ceratopsids; P = pachycephalosaurids; S = stegosaurids; N = nodosaurids; A = ankylosaurids. (*Based in part on data from Norman, 1985*)

165 million years of geological and meteorological history as seen through a wide-angle lens, so to speak. But when we increase the resolution, trying to capture the age of dinosaurs in finer detail, the picture that emerges is a great deal more complex and problematic. Most problematic of all, from the viewpoint of paleontology, is that the fossil record is incomplete. Large parts of the Mesozoic world

have disappeared without leaving behind a clear trace, not so much as one iota of direct evidence to indicate which, if any, dinosaurs lived in certain places at certain times.

As you probably know, fossilization can occur only if a number of strict requirements are met. For one thing, the organism has to die where burial will occur rapidly enough to prevent decay and weathering. Even teeth, horns, and bone, when exposed to wind, rain, and temperature extremes, will decompose beyond all recognition. For another, the means of burial must be gentle enough to avoid crushing and disintegration. This is why most fossils are found in marine sedimentary rocks. When aquatic organisms die they sink to the bottom, where they are soon covered in a protective layer of mud or other fine-grained sediment settling out of the water. Over time, under their own weight, the cumulative layers of sediment will become compacted and cemented together to form sedimentary rocks, shale and limestone primarily, while the hard parts of the organisms buried within the layers are either preserved more or less as they are (shells, typically) or are replaced by dissolved minerals (a more likely outcome in the case of bones). Throughout the entire process—death, burial, compaction, and, should it occur, remineralization—the sediments cannot be disturbed. No doubt lots of plants and animals have been buried in the geological environments that are home to igneous and metamorphic rocks, but exceedingly few fossils have survived the violent processes by which such rocks are formed.

What the selectivity of the geological record means for those of us who study fossils will be easier to grasp if you look closely at a typical present-day depositional environment—the Gallatin Valley, for example, in southwestern Montana, where the Museum of the Rockies is located. Sediment is being deposited in the valley all the time, as gravel along streams, sand in floodplains, silt and clay at the bottom of ponds, reservoirs, and lakes; and most of the sediment comes from the mountains that surround the valley or are located upstream of it—the Gallatin, Bridger, and Madison ranges, in particular. Now, jump ahead in time ten million years. What has become of the depositional environment that we once knew as Gallatin Valley?

Most notably, the mountains are gone, completely and irreversibly eroded away. Large-scale geological structures, you see, do not show up in the geological record as large-scale structures but instead as strata of sediment or sedimentary rock. Small variations in the original landscape, streambanks and shorelines, for instance, might be preserved, but certainly nothing larger. And if the hills and mountains are erased, so too are any plants or animals that might have lived in the hills and mountains. True, bones are sometimes carried long distances by streams and deposited in sand or mud, where they are buried and fossilized, but like characters without stories, such displaced specimens can tell us little more about their lives than that they once existed somewhere, sometime, somehow. Actually, the situation is worse than that, because whole skeletons cannot survive movement of any kind, much less a long, rough journey by water; when we find fossils that have been transported from another location after death they are always fragments, jumbled together at random, and often damaged.

The paleontologist who ten million years hence excavates the area that used to be the Gallatin Valley will, if she is very lucky, find fossils of only those animals that lived and died in the valley. No amount of luck will turn up the remains of a mountain goat, however, the goat having gone the way of the mountain. But, you might argue, we can infer that there once was a mountain, so why can't we infer that there was a mountain goat? A good question but, all the same, based on a misunderstanding. The only reason we know there once was a mountain is because we have witnessed erosion; we have seen with our own eyes where sedimentary deposits come from—mountains—allowing us to reason backward, using current sediments, along with other clues, to reconstruct a physical environment that existed a long time ago. One of the cardinal principles of geology is something called uniformitarianism, which means simply that the geological processes of the past are the same processes we witness today. And remember, too, that the goat in the example represents an animal for which we have no other evidence than what might show up as a fossil. The biological processes of the past are the very same processes that operate today, but if the dinosaurs teach us anything, it is that the organisms that arise from those processes can vary greatly from one period to another. If there existed high-altitude dinosaurs during the Mesozoic era, dinosaurs that never left the mountains, our chances of finding their remains today are slim, and of being able to understand such rare remains even slimmer.

Okay. Let's complicate things a little further, not to be mischievous but to suggest something of the actual difficulties one encounters when trying to plumb the mysteries of the natural world. Let's say that a paleontologist returns to the area once called the Gallatin Valley a hundred million years from now. What will she find? Among the many entirely normal events that might have occurred in the meantime is an overall depression of the central part of North America, causing eastern Montana to tilt downward. This, in turn, would have accelerated stream flow and the rate of erosion east of the Continental Divide, where the Gallatin Valley is located. All along the Missouri River today there are cascades and waterfalls. If the river ran faster, the cascades would grow more pronounced and the waterfalls would migrate upstream as the water, rushing at a furious pace, ate away the bottom. Cascades of various size would also form in the tributaries of the Missouri—the Gallatin, Jefferson, and Madison rivers—and they, too, would edge ever farther upstream, toward their headwaters, eventually eroding out all of the sediments that were deposited in the Gallatin Valley region, erasing even the pulverized vestiges of the mountains that once encircled the valley. Anyone visiting the region a hundred million years from now would find no record of what occurred immediately prior to erosion—that is, no sign of life as it is now, neither the original sedimentary rocks nor the fossils they might have contained.

Piecing together stories a hundred million years old or older on the basis of incomplete and altered geological records is exactly what occupies paleontologists today. And what I've described so far represents only a few of the factors that must be taken into consideration when trying to do so. As in the case of fossilization in marine sediments, some depositional environments are more likely to preserve the remains of animals than others. Even where the terrain is comparatively flat, the composition of sedimentary rock can vary significantly from one region to another. Upland areas, for example, which are those that lie closest to mountains, are usually well drained and thus relatively dry. They comprise an abundance of mudstones but very little sandstone, and tend to be green or red in color, which indicates that they contain large quantities of sodium, potassium, and other alkaline chemicals. And as it happens, upland alkaline sediments preserve calcium-based bone better than they do carbon-based plant material. Lowlands, by contrast, being located near seas and lakes, are poorly drained. They tend to be swampy and acidic, rich in hydrogen and generally made up of more sandstone than mudstone. Tan, gray, and sometimes black, acidic lowland sediments preserve plants better than bone. Coal, for example, which is composed largely of carbonized leaves and stems, is a typical lowland sedimentary rock. A good example of a contemporary upland area is the eastern two-thirds of Montana, which, because of its proximity to the Rockies, is high and arid, whereas Louisiana, along the Gulf of Mexico, and, say, New Jersey, near the Atlantic Ocean, possess all of the characteristics of lowland environments.

Why is it important to think about this? I can tell you why it's important to me. I want to find dinosaur eggs and baby dinosaur bones, among other things, and I want to find them as quickly and efficiently as possible. I could devote years to searching ancient swamp environments without finding a single one, not because dinosaurs never inhabited such areas but because the acidic sediments might have long ago dissolved whatever bones and teeth and horns the dinosaurs left behind, obliterating their remains forever. Finding dinosaurs, in other words, is as much a matter of knowing where evidence is likely to have been erased as where it might still be preserved. This is one of the reasons why I have spent most of my professional life hunting for fossils in central Montana. A great many of the geological formations that happen to be exposed there represent upland regions from the age of dinosaurs, more precisely, the Cretaceous period, when the two groups that have interested me the greatest during the past ten years—the duck-billed and horned dinosaurs—inhabited the plains along the Rocky Mountain Front from Alaska to Mexico.

It's time, I think, to focus our historical lens on North America between 136 and 65 million years ago and increase the magnification, bringing Montana into sharper view. Across the world, remember, continents are being driven apart from each other and new oceans are forming in the rift zones widening between them. The global climate is considerably warmer and more humid than today. There is no ice on Earth, not even at the South Pole, where, then as now, Antarctica is located. Remember, too, that large parts of North America lie at low elevations relative to the newly formed oceans. A vast body of water, the Western Interior Cretaceous Seaway, has flooded the central and southern reaches of the continent and now stretches from the Gulf of Mexico to the Arctic, isolating the Rockies from the rest of North America. The seaway is fed by tropical waters from the Gulf, and it's shallow, which means that it's also solar-heated. Crocodiles live comfortably along its shores in what is now northern Alberta. But most important, the seaway is dynamic, rising and falling at least three times during the Cretaceous, on each occasion expanding its boundary significantly westward, toward the Rockies.

Let's increase the magnification again. Uplifted during the Triassic period, the Appalachians in the east are pretty old and worn down now; little terrestrial sediment is being deposited anywhere in that region. On the other side of the continent, however, the Rocky Mountains are newly formed. They thrust above the continental floor and, significantly, they thrust eastward, which has created a broad depression, similar to a rumpled rug, which geologists call a foredeep, all along the front. Immense quantities of ter-

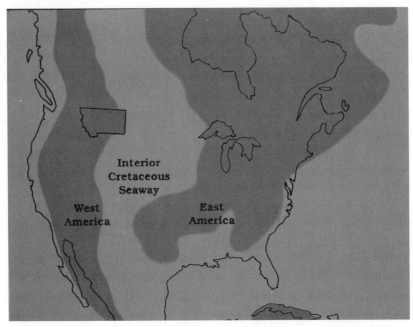

North America, showing the state of Montana and the position of the Interior Seaway during the Cretaceous period.

restrial sediments formed during the erosion of the mountains are not only carried eastward by creeks and rivers and, to a lesser extent, wind, they tend to be deposited in the foredeep, migrating no farther. In succeeding chapters I'll fill in the details of this picture, especially as they concern the evolution of duck-billed and horned dinosaurs in the late Cretaceous, but for the time being it is sufficient to appreciate in very general terms how the erosion of the Rockies and the rise and fall of the Western Interior Seaway interacted to produce the sedimentary formations found in central Montana today. The two depositional processes in effect dovetailed, three fingers of marine sediments, representing three expansions of the seaway, interleaved with four fingers of terrestrial sediments, representing the seaway's contractions.

As you can see on the stylized cross-section chart on the following page, each distinct marine and terrestrial formation bears

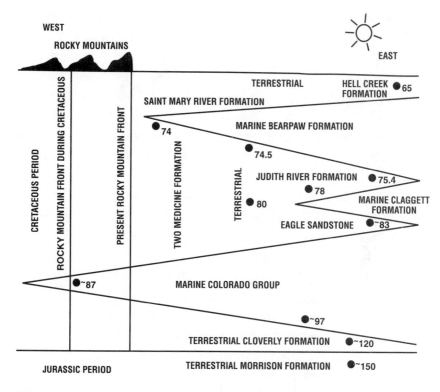

WEST

ROCKY MOUNTAINS

EAST

TERRESTRIAL

SAINT MARY RIVER FORMATION

HELL CREEK
FORMATION ● 65

ROCKY MOUNTAIN FRONT DURING CRETACEOUS

PRESENT ROCKY MOUNTAIN FRONT

●74

MARINE BEARPAW FORMATION

TWO MEDICINE FORMATION

●74.5

CRETACEOUS PERIOD

JUDITH RIVER FORMATION ● 75.4

TERRESTRIAL

● 78

● 80

MARINE CLAGGETT
FORMATION

EAGLE SANDSTONE ●~83

●~87

MARINE COLORADO GROUP

●~97

TERRESTRIAL CLOVERLY FORMATION ●~120

JURASSIC PERIOD

TERRESTRIAL MORRISON FORMATION ●~150

This cross section of the dinosaur-bearing sediments of central
Montana shows how the marine sediments interfinger from the east.
The numbers are the relative age-dates of the different strata.

its own name, most of which you needn't be concerned about
right now. But I want to call your attention to the Judith River
Formation, and especially to the time between 75.4 million years
ago, when the seaway had regressed to its easternmost point, and 74
million years ago, when the last transgression onto the plains came
to a halt and the seaway started to retreat again. It is in this suite of
terrestrial sedimentary rocks, you may recall, that we found the
splendid *Brachylophosaurus* skeleton. When earlier I said that I was
seeing that particular duck-billed dinosaur in its actual historical
context, in terms of the environment in which it lived, I was referring in part to the information represented in this chart, information

that resulted from geologists painstakingly dating the sedimentary rocks of Montana and mapping their extent, location, and stratigraphic relationship to one another. In the chapters to come you will be hearing more about the Judith River Formation, but the group of sedimentary rocks that has intrigued me most lies to the west—the Two Medicine Formation. The Willow Creek anticline, where we first discovered eggs, nests, and babies, is located in the Two Medicine, as are many of the sites where we have excavated dinosaur fossils during the ten or so years that have passed since leaving the anticline and Egg Mountain. (Strictly speaking, we haven't left the anticline; the Museum of the Rockies still operates a field school there and every summer students find more eggs, nests, and baby dinosaurs.)

Why focus on the Two Medicine Formation? The fossil record, we know, is both biased and incomplete. Dinosaurs lived all over the world, but certain of their native environments have disappeared, as mountains do during erosion, along with any remains that might otherwise have survived, or they possess some characteristic that reduced the chances of preservation, such as the acidic quality of lowland swamps. The Two Medicine Formation, however, represents the upland reaches of a coastal plain that at its widest point, when the Western Interior Cretaceous Seaway had fully receded, was four hundred or more miles across. Over a period of about 12 million years, from 84 to 72 million years ago, streams poured out of the young Rocky Mountains, routinely and extensively flooding the plain and leaving in their wake layer upon layer of sand, mud, and other sediment, which eventually turned into rock. Today, in north-central Montana, the cumulative layers of sedimentary rock that make up the Two Medicine Formation are two thousand feet thick. The surface of the formation covers thirty-six hundred square miles and extends from extreme southern Alberta to Augusta, Montana.

The formation needn't have been that thick or extensive. If the original deposition had occurred more slowly, that same 12 million years might be represented by one thousand feet, or five hundred, or fifty. Even the most comprehensive of geological columns contains gaps, places in the strata of rock where deposition ceased altogether or some kind of secondary erosion took place, erasing sediments that had already been deposited. Such gaps are particularly troublesome because there is no telling how long the pause in deposition or the erosion event lasted, whether a hundred years, a thousand, or a million. A gap in the geological record reveals only that something is missing; it says nothing whatsoever about what is missing, or how long a period the gap represents. And in truth, comprehensive records are rare, the intervals in most geological columns far outnumbering the surviving sediments. Deposition is simply too variable and erosion too universal to permit much else.

You can now better appreciate why I like north-central Montana so much. The sedimentary rock there has been uplifted, folded, eroded away in places, even, at times, twisted completely out of shape, but the strata as a whole remain close enough to their original orientation and condition to be deciphered and compared with other strata, within the Two Medicine Formation as well as those of other formations. What's more, the thickness of the Two Medicine Formation allows me to look at a relatively long period of time in the natural history of dinosaurs in some detail. In other words, the "resolving power" afforded by the sedimentary strata is strong.

Critical to any understanding of evolution is the ability to see relationships among organisms that lived at the same time—in what ways they are similar or dissimilar—as well as relationships among organisms that lived at different periods, that is, how certain characteristics may have changed over time. In many instances, the tools, both physical and conceptual, that we have used in the past to study evolution have lacked sufficient resolving power to bring these relationships into view. Were I to concentrate my fieldwork in a formation that is only five hundred feet thick, for example, the behavioral and evolutionary information I seek would remain fuzzy, out of focus, if they could be detected to any useful degree at all. The best dinosaur stories—that is, the least ambiguous and most clearly defined stories—come from the thickest, most complete, least disturbed sediments. In the section of the Two Medicine Formation that surfaces in north-central Montana, the same amount of time—twelve million years—is represented by four times as much rock as my hypothetical five-hundred-foot formation. This means that if a dinosaur died in a floodplain along the Rocky Mountain Front, it was four times more likely to be buried in sediment and thus fossilized. The geological column captures more detail.

Allow me one final observation about the Two Medicine Formation, and about dinosaur hunting in Montana overall, before I describe recent excavations. My primary interest is the Cretaceous period, especially the late Cretaceous, from about 80 million years ago onward, because that is when the ornithopods, or duck-billed dinosaurs, such as *Maiasaura* and *Brachylophosaurus*, flourished, as did the ceratopsians, or horned dinosaurs, the group that includes *Protoceratops* and *Triceratops*. Both groups appeared late on the dinosaurian tree, the ceratopsians very late, and they were among the last of the dinosaurs to walk the earth. More important from my standpoint is the fact that both groups formed herds. Other dinosaurs may have gathered together or acted in concert for one reason or another, but the only groups we can be certain did so are the duckbills and their closest ancestors, as well as the horned dinosaurs. And the questions that drive most of my research, regarding social behavior and the evolution of species from one generation to another, can only be addressed by comparing large numbers of the same animal. The Two Medicine Formation of north-central Montana fulfills that requirement. Indeed, with its detailed and comprehensive record of the comings and goings of immense populations, it is the Serengeti of dinosaur deposits.

My first interests are the ornithopods and ceratopsians, but they are not my only interests. In the fieldwork section you'll find that the cast of characters includes as well many other kinds of dinosaurs, most prominent among them certain key theropods (*Tyrannosaurus, Allosaurus, Deinonychus*) and sauropods (*Apatosaurus,* previously called *Brontosaurus*). You'll also see that we have not confined our explorations to the Two Medicine Formation or to upland environments or even to the Cretaceous period. Montana as a whole is fossil country, especially the eastern two-thirds of the state, from the Rockies well into the Great Plains, and I try to take advantage of all of the opportunities it offers. In this I am truly walking in the footsteps of the pioneers of paleontology, who found in Montana a trove of dinosaur fossils. In 1856, Ferdinand Vandiveer Hayden, a geologist, was the first to discover dinosaur remains in North America—a variety of teeth—near the confluence of the Judith and Missouri rivers, in the central part of the state. Some of the teeth belonged to a duckbill from the late Cretaceous that came to be known as *Trachodon;* others to the man-size theropod *Troödon;* still others to unspecified horned dinosaurs and a tyrannosaur named *Deinodon.*

Almost anywhere you search for dinosaurs in Montana you find yourself immersed not only in geological history but the history of American scientific exploration. Near Billings, at the turn of the century, Earl Douglass found numerous duckbill fossils. Some of the specimens, I learned when I examined them at Princeton University in the late 1970s, were juveniles, and that was the revelation that inspired my search for baby dinosaurs. Over the course of several expeditions to Montana between 1902 and 1916, one of the most successful dinosaur hunters of all time, Barnum Brown of the American Museum of Natural History, collected untold numbers of specimens, including *Triceratops* bones and the world's first *Tyrannosaurus rex* skeleton near Jordan, in the eastern part of the state, and duckbill skeletons from an area near the Two Medicine River, in the north-central part. And as we saw in chapter 2, John Ostrom's analysis of *Deinonychus,* from south-central Montana, was the first serious challenge to the conventional idea that all dinosaurs were cold-blooded, sluggish reptiles.

But in the spring of 1985, when I was casting about for new places to dig up dinosaurs, the region that most appealed to me was the Landslide Butte badlands, in extreme north-central Montana. In Gilmore's field diary from 1928, twelve years after he had used what he mistakenly thought was a layer of clamshell fragments as a datum, or benchmark, in the local sedimentary column, he described collecting dinosaur eggshell fragments. By that time he knew what he was looking at and he said so in his notes, clearly and unequivocally, but for reasons I cannot fathom he never published the find, despite the worldwide excitement the discovery of eggs in Mongolia had stirred up just a few years earlier. Be that as it may, I wanted to study more eggs and more juveniles, and Gilmore had found both. In addition, he had found them in the Two Medicine Formation, the same rock in which the Willow Creek anticline was located. Reluctant as I was to leave the anticline, Landslide Butte seemed like a promising alternative. What I didn't realize at the time, however, is that it was more than that, much more.

4

BEYOND EGG MOUNTAIN

What would it be this time? An outcrop of rock that had weathered just so, its shape suggestive of some mysterious creature? Or the not-very-old skeleton of an all-too-familiar animal that for some reason hadn't been recognized as such? These were the questions that were rolling through my mind as Bob Makela and I drove to the Blackfeet Indian Reservation in north-central Montana, a journey of some 250 miles from my home in Bozeman.

It was the fall of 1984, the year we removed thirty tons of rock from Egg Mountain. I'd received a call from Marvin Weatherwax, a member of the Blackfeet Indian Tribal Council. A friend of Marvin's, it seems, had spotted a dinosaur skeleton, or what he thought might be a dinosaur skeleton, lying on the bottom of Four Horns Lake, in a shallow area just offshore. Marvin's description of the location and orientation of the skeleton led me to believe that the animal possibly had beached itself and consequently died—that is, if it wasn't a dinosaur but something of more recent, less exotic origin.

But there was no way of knowing this or anything else without actually traveling to the site and looking for myself. And that's where the misgivings came in. Since digging up our first dinosaur skeleton together in the early 1970s, Bob and I had been on many wild goose chases, traveling from one end of Montana to the other—and Montana is a huge state—only to find that what had been described to us as a clutch of dinosaur eggs was really an accidental grouping of small, water-worn boulders or that the well-preserved *Triceratops* specimen that had so excited our informant was well preserved, all right, but belonged instead to a woolly mammoth only twenty-five thousand years old, interesting on its own terms, certainly, but hardly a dinosaur. Still, we continued looking, following the most promising leads, because, on occasion, curious individuals and amateur collectors make important paleontological discoveries. And sometimes it happens when we least expect it.

Bob and I certainly weren't expecting much as we drove the last few miles to Four Horns Lake. In fact, what we saw only served to increase our skepticism. There was no sign of Mesozoic rock anywhere. Every roadcut we passed was comprised of glacial till—loose stones ranging in size from gravel to boulders—that had been deposited during the last Ice Age, about ten thousand years ago. Bob shook his head. "Here we go again," he said. I wondered how long it would take to get back to Bozeman.

But the approach was misleading. As it turns out, Four Horns Lake rests within a recess in a small outcrop of the Bearpaw Shale, marine sediment that was deposited during the last expansion and contraction of the Western Interior Cretaceous Seaway, between 75.4 and 65 million years ago. Even more surprising, Marvin Weatherwax's account had not been that far off the mark. Although the skeleton his friend had found was not that of a dinosaur, the creature it did belong to was every bit as old and almost as interesting—a mosasaur, the very same animal that Georges Cuvier had first identified two hundred years ago, thereby providing some of the earliest persuasive evidence for the idea of extinction. *Mosasaurus* lived throughout the Cretaceous period and, along with other marine reptiles, dominated the oceans of the world just as the dinosaurs dominated the land. The animal's head resembled that of a crocodile, it had four flippers and a powerful paddlelike tail, and it grew to thirty feet. Marvin's mosasaur was about twenty feet long and lay in water three feet deep, ten feet from shore. Judging from the shale in which the fossil skeleton was embedded, I estimated its age at about 73 million years. Most of the bones were intact. And, as I say, submerged. Bob and I decided to postpone the excavation till the following spring, before the runoff began, when the water would drop to its lowest level of the year. And we surely would return. In Kansas mosasaurs are common, which is not surprising, since the sediments there represent the middle of the Interior Seaway. To find one in Montana, in rock representing the westernmost edge of the seaway, is highly unusual.

In April, when we returned to Four Horns Lake, the weather was nasty, winter clinging desperately to the land long after its charms had worn off, an unpleasant phenomenon that is fairly common in Montana and thus unlikely to discourage outdoor activity, least of all among natives of Bob Makela's temperament. To Bob, the sleet and freezing wind were inconsequential compared with the satisfaction he hoped to gain by excavating a specimen immersed in thick, sticky mud and a couple of feet of murky water. I've never known anyone who enjoyed tackling practical problems as much as Bob did, nor who was more capable of doing so. If you ever crashed in the snowbound Andes or found yourself stranded without food or water on a volcanic island, Bob was the guy you would want as your partner.

The problem at Four Horns Lake was nothing like trying to survive in a hostile environment, of course, but challenging nonetheless: finding a way to make a hole in the water. Bob's solution was to construct a swimming pool over the specimen, then drain it. He, I, and a small crew pounded sharp-tipped steel fence posts into the bottom of the lake, then attached thick sheets of plastic to the posts, completely surrounding the site. We then set up a sump pump to remove water from within the makeshift dam. That's when we dis-

Our attempt to drain a portion of Four Horns Lake to collect a mosasaur skeleton.

covered that the mosasaur skeleton was encased in large nodules of marine limestone that had formed within the layers of shale. Many of the bones were encased in their own nodules, however. Excavating them was simply a matter of mapping their location and carrying them to shore. Other, more fragile sections we covered with plaster jackets, a practice we usually follow to prevent breakage when we remove fossils from the ground and carry them elsewhere. It took several long, wet days to transfer the specimen from the lake to the back of my truck.

The mosasaur now rests on a shelf in the basement of the Museum of the Rockies, awaiting preparation, along with more jacketed specimens than I care to contemplate at the moment. What I couldn't have anticipated back when we excavated that primitive seagoing reptile, and what I sometimes marvel at now, is the profound effect it eventually had on the direction my research took during the rest of the 1980s. I certainly believe in the adage that chance favors the prepared mind, but even the best prepared are unlikely to go far in the complete absence of such favors, especially in a non-mathematical, historical science like paleontology, where fieldwork still serves as the foundation of all that we do. The paleontologist never knows what awaits him over the next hill or during the next season, nor toward what new horizons those discoveries may show the way. In 1978, Bob and I stopped at a rock shop in the tiny town of Bynum, Montana, north of Choteau, motivated by nothing more than curiosity and a tip from our Berkeley paleontology friends. Seeing that we knew something about fossils, the owner, Marion Brandvold, asked us to identify two bones she had stored away. She then handed me the thighbone and jaw of a baby duck-billed dinosaur, which she said she had collected on Jim Peebles's cattle ranch near Choteau—the Willow Creek anticline. I had found what I wanted most to find in the last place I expected to find it. In 1984, another happy accident happened: Marvin Weatherwax's friend chanced upon a funny looking skeleton in the bottom of Four Horns Lake. And promiscuous curiosity outweighing hard-earned skepticism, as always, Bob and I visited the lake, an event that would be the key to gaining access to an area I had long wanted to excavate. Not right away, though. Fieldwork prospects were in fact looking bleak that inclement spring.

Just about the time that we were excavating the mosasaur, I received another call, this time from Jim Peebles, who informed me that, as much as they hated to do do it, his family could no longer allow my crews onto the Willow Creek anticline. I was shocked. Fourteen students and volunteers had already signed up for the 1985 summer season. Supplies and equipment had been readied. Worst of all, I didn't know where else I might turn to find more eggs, nests, and babies. At the same time, however, I appreciated the Peebles family's predicament. Word of our discoveries had spread. Amateur collectors routinely trespassed on their land, growing ever bolder in their unauthorized search for the fossil treasures of the now-famous site that Bob had named Egg Mountain. The situation had grown unmanageable, the probability of an accident or unfortunate encounter increasing daily. After five years of putting up with me, Bob, and our crews traipsing all over their land, they simply didn't have the patience to deal with the crowds arriving at their front door. Who can blame them?

That said, my concerns now lay elsewhere: I had less than three months to find another site. The most reasonable approach, given the little time available, was to identify areas that met the criteria we had determined were critical to our success at the Willow Creek anticline, which had taught us much about the habitats of the two ornithopods we had found there—the new duck-billed dinosaur, *Maiasaura,* and the smaller, fleet-footed *Orodromeus,* a new hypsilophodon—and the theropod *Troödon.* The most promising new locations would be the sedimentary remnants of upland coastal plains from the Cretaceous period. The sediments would consist primarily of red and green mudstone, like those at the anticline, and contain caliche, limestone nodules that form beneath the surface as groundwater fluctuates, depositing calcium carbonate each time it rises. The location of caliche nodules indicates how wet a climate is: the more rainfall, the nearer to the surface the nodules of calcium carbonate accrete. The size of the nodule, on the other hand, reflects the length of time that the soil above it remains stable. If the surface level changes (through erosion, say), the elevation of underlying groundwater shifts and a new caliche layer forms. Caliche layers provide a surprisingly precise way to determine whether different groups of fossils were deposited on the same soil surface, or horizon—that is, at the same time, a crucial element in the study of the social behavior of dinosaurs.

A few weeks later, in May, Mick Hager, who was director of the Museum of the Rockies at the time, and I were invited to a site called Devil's Pocket near Harlowton, in central Montana, that met my criteria. Though from the air Devil's Pocket looks for all the world like a meteor crater, it is in fact a different though relatively common structure called a dome—a rock formation, in this case, layers of sedimentary rock, that has bubbled up on the surface under pressure from rising volcanic magma deep within the earth's crust. The dome at Devil's Pocket formed so long ago that its center has been eroded away, producing a bowl-like feature about a half-mile wide. Exposed along the rim of the bowl are walls of red and green mudstone containing caliche. The sediments are 110 million years old, deposited early in the Cretaceous period, and part of an upland area of the coastal plains called the Cloverly Formation.

John Ostrom had explored Devil's Pocket in the 1960s and found nothing, having arrived, evidently, at the wrong time in the erosion cycle, a slow process in central Montana, where precipitation is scarce and decades pass before new fossils are exposed at the surface. But thirty years earlier Barnum Brown, whom chance always seemed to favor, discovered and excavated the remains of a *Sauropelta*, an ankylosaur, or armored dinosaur. *Sauropelta* lived at the same time and in the same general environment as *Deinonychus* and *Tenontosaurus*, a primitive, plant-eating ornithopod upon which *Deinonychus* preyed. That *Tenontosaurus* belongs to the same suborder as *Maiasaura* gave me reason to believe that it could have nested in colonies and cared for its young, just as *Maiasaura* did. Devil's Pocket, I thought, might contain eggs and babies.

On a scouting trip late in May, our host, Don Rassmussen, an oil geologist from Denver who had grown up in Harlowton, happened upon a microsite, which is a collection of small fossils, usually sorted according to size but representing different animals, that were washed together by a stream after the animals died. Weathering out of the mudstone at Don's microsite were numerous teeth—from *Deinonychus*, a sauropod of unknown identity, a mammal, and a lungfish—along with a variety of snail shells. Mick, for his part, reenacted Brown's discovery, finding a partial *Sauropelta* skeleton. I had the good fortune of collecting a single *Tenontosaurus* toe bone. Mildly encouraged, I sent Bob and his crew to Devil's Pocket on June 12. During the next two weeks they dug up Mick's *Sauropelta*, part of a sauropod that Don had found during a previous visit, another sauropod's tail, and the partial skeleton of a juvenile *Tenontosaurus*. I collected some eggshell shards, the first ever from the Cloverly Formation, but too small and too few to identify. There was little else to be found. As tantalyzing as the Cloverly was proving to be, erosion would have to scour the rim of Devil's Pocket for a long time to come before additional fossils were exposed. I still hadn't found a suitable alternative to the Willow Creek anticline.

The previous summer Jill Peterson, a student at the University of Colorado, and I had conducted a preliminary investigation of a site south of Choteau called Red Rocks. Like the sediments of the Willow Creek anticline, which are 76.7 million years old, Red Rocks is part of the Two Medicine Formation but from older, lower strata in the geological column, dating to about 80 million years ago. Jill and I had found a protoceratopsian skeleton, baby bones, and a handful of eggs. Perhaps there was more. Toward the end of June, while Bob and his crew finished their work at Devil's Pocket, I sent another, smaller crew to Red Rocks. They found lots of unidentifiable eggshells and baby bones, as well as a duck-billed dinosaur and a horned dinosaur. But Red Rocks, I soon realized, resembled Devil's Pocket, in that it was a relatively small outcrop in which few fossils had been exposed. I was glad to learn, however, that eggs and babies could be found in the lower half of the Two Medicine Formation. Charles Gilmore and others had insisted that only the upper half contained dinosaur remains.

Just like that, June was over. Frustration soon set in, since it was beginning to look as if we were going to spend the entire 1985 season wandering from one disappointing dig to another. My thoughts returned to Gilmore's field diaries from Landslide Butte. In addition to the two new species of horned dinosaur, *Brachyceratops montanus* and *Styracosaurus ovatus*, he had collected juvenile dinosaurs and eggshell fragments. He even mentioned having quit excavating one quarry because it contained several skeletons of a single species, all of them the same size—small, which he interpreted as the adult form of a small duck-billed dinosaur. But I'd seen the skeletons; they are instead the juvenile form of a duck-billed dinosaur of average size—in other words, a group of young dinosaurs, exactly what interested me most. Apparently Gilmore, like his contemporaries and many collectors today, would have considered further digging worthwhile only if it promised to yield a single skeleton of a large individual. As for me, I'd much rather devote a week to collecting dozens of specimens that fit into Ziploc bags and provide clues to dinosaur social behavior and evolution than a year to excavating a single specimen that tells me little more than that I now have in my possession a skeleton too large to house at the Museum of the Rockies.

Convinced by Gilmore's discoveries and intriguing misinterpretations—remember, paleontology often thrives on both—that the Landslide Butte badlands held much promise, I had approached the Blackfeet Indian Tribal Council on several occasions, starting as early as 1982, seeking permission to collect fossils in the extreme northern region of the reservation. Each time the council had turned down my request, in the belief that our activities, the comings and goings of field crews, the exploration trips, and the excavations themselves, would damage the land or disrupt the lives of reservation residents. Early in July, desperate to salvage what was left of the 1985 season, I went before the tribal council once again. This time, however, I found support—from Marvin Weatherwax. He had witnessed the excavation Bob and I had completed at Four Horns Lake only a few months earlier. He had seen how little land we actually disturbed during the operation. And he persuaded his fellow council members that we could be trusted. They then granted us permission to explore and collect not only in the area near Landslide Butte, to which I had limited my request, but on the entire reservation. Like a 200-million-year-old marine version of a good luck charm, the mosasaur had unexpectedly opened the door to one of the richest fossil deposits in the world.

The six-mile expanse of badlands that lies between Landslide Butte and the Canadian border represents the top three hundred feet of the Two Medicine Formation, the uppermost layers of which are about 74 million years old. The strata consist of, among other things, silty, caliche-laden mudstone, red beds as well as green beds, all of which are the remnants of extensive floodplain deposits along the eastern front of the Rockies. Resting immediately above the Two Medicine Formation is the Bearpaw Shale, the same unit of marine sediments from which Bob and I had recovered the mosasaur. The shale is widespread, too, allowing us to make precise measurements of the distance between the top of the Two Medicine Formation, which the shale always defines and which we therefore adopted as our datum layer, and any dinosaur remains located at lower, older levels of the strata. And with that information we were able to determine the age of particular dinosaur specimens, and whether dinosaurs removed from different sites lived at the same time.

As you can see from the cross section back on page 47, the Bearpaw Shale was deposited during the last transgression of the Western Interior Cretaceous Seaway. Try to bring into focus again the implications of this event. About 75.4 million years ago, the coastal plain is as wide as it has ever been—maybe as much as four hundred miles from the foot of the still-growing mountains to the edge of the sea. The water starts to rise, the plain starts to narrow. It is a slow process, exceedingly slow by any human reckoning, but immensely consequential for the plants and animals that live on land, including the dinosaurs—ornithopods, ceratopsians, theropods, ankylosaurs—that now dominate most terrestrial habitats. By the time the sediments of the upper Two Medicine Formation are deposited, 74 million years ago, the coastal plain has shrunk to one-eighth or less of its original size. It is only thirty to fifty miles wide now, and into that much smaller space all of the dinosaurs are corralled—predator and prey, enormous herds of duck-billed dinosaurs and horned dinosaurs, and the cousins of *Tyrannosaurus*, which, we can be certain, never wanted for a carcass to scavenge. Etched into the badlands of Landslide Butte is a record of this critical time in the evolutionary saga of dinosaurs.

Another picture that you might keep in mind, and which will help you make sense of the chart, is that as the seaway advanced westward, toward the Rockies, silt and other fine-grained material settled to the bottom, directly atop the sand and mud that had been deposited on land by streams and rivers flowing out of the mountains, and in time those marine deposits turned to rock—the Bearpaw Shale—as did the terrestrial deposits, which, of course, are what we now call the Two Medicine Formation. As time passed and the seaway continued expanding, more and more terrestrial deposits were buried beneath marine deposits. All of this occurred tens of millions of years ago, and geological processes being what they are, what we actually see today is not nearly this simple. Indeed, if you think about it, were it only a matter of the orderly deposition of sediments, what we would see today would consist solely of the most recent deposits—the rock we call the St. Mary River Formation along what was the upper part of the coastal plain and the Hell Creek Formation along what was the lower. Any marine deposits would be hidden deep underground.

Land surfaces, of course, are subject to a range of forces, some extremely violent, that alter their structure and composition. The sedimentary rock along the Rocky Mountain Front was uplifted and tilted, folded and fractured. Erosion occurred at different rates in different places and exercised different effects on different types of sediment. Some areas were buried in volcanic ash. Others were carved by glaciers. As a consequence, traveling west to east today or, for that matter, north to south, you may encounter rocks of vastly different age and makeup sitting side by side. In some cases, the rocks have been so jumbled up that it's nearly impossible to tell one from the other. Sometimes an outcrop of one formation will rest like an island within a sea comprised of the rock of an entirely different formation. A good example is the section of Bearpaw Shale that forms the bottom of Four Horns Lake. Among the most prominent geological features of north-central Montana today is the Sweetgrass Arch—a dome, like the one at Devil's Pocket but larger and more pronounced, and which extends east from Choteau about sixty miles. Along this bulge all the terrestrial sediments of the coastal plain have been eroded away, exposing the Colorado Shale, which was deposited during the first expansion of the seaway and thus contains countless marine fossils. Because the Sweetgrass Arch divides the plain in two, the lowlands have been given a different name, the Judith River Formation, than the uplands, the Two Medicine.

After the tribal council granted us access to the Blackfeet Reservation, we next had to secure the permission of the individuals who live in the areas where we wished to collect. One of these was Ricky Reagan, an Indian who ranches along the Milk River in the middle of the Landslide Butte badlands. Ricky is a rodeo cowboy and likes to brand calves the old-fashioned way—by chasing them down on horseback and lassoing them with a rope. He also puts off branding a few weeks longer than most Montana ranchers, allowing the calves to grow larger, stronger, which makes them more fun to wrestle. That's what Ricky and his neighbor Arne Johnson were doing when I arrived to talk to him about hunting for dinosaur fossils on his property. Although clearly valuing the ranching life as much for the privacy it offers as anything else, Ricky generously consented, opening his land to a scouting party of that ragtag band known as paleontologists, which was a relief to me. At long last, I would be able to explore the area that no member of my profession had seen since Gilmore visited decades before.

The expedition began on a bright note. Early in August, while scouting for a place along the Milk River to set up camp, Bob came across a promising bone bed, although, at the time preoccupied with matters of access and comfort, he promptly forgot its exact location. Several days later, Jill Peterson, my son Jason, who was thirteen at the time, and I rediscovered the site—a layer of sandstone about three feet thick that capped a small butte. Dino Ridge Quarry, as we called the area, contained the remains of what turned out to be a new species of horned dinosaur that would eventually be named *Einiosaurus*. A late Cretaceous ceratopsian that grew to eighteen feet in length, *Einiosaurus* weighed about two and a half tons and possessed a single, imposing nose horn. From its neck shield two long spikes were arrayed. Our efforts to excavate the site were stymied, however, when the weather turned foul, then stayed that way—for weeks. My field notes covering the rest of August read like an advertisement for an umbrella company. Rain followed by relentless drizzle interrupted now and again by downpours. The sun abandoned us altogether. Sky and earth alike turned to mud, distinguishable from each other only by their color: wet-cement gray and saturated brown. Chill, damp winds howled through the gullies and outcrops of the badlands. If there was a dry piece of clothing in camp I didn't see it, and I certainly wasn't wearing it.

In the midst of these discouraging conditions, however, we managed to discover a second important bone bed. On an afternoon when I simply couldn't sit still any longer, I trudged out into the rain looking for something, anything, that might justify our continued presence at Landslide Butte. The muddy ground rendered hillsides inaccessible, so my search was confined to the flats near camp. To my amazement, only two hundred yards from where I had been stewing all day, I found, among other remains, an adult thighbone of a new species of *Hypacrosaurus,* a crested duck-billed dinosaur that in its adult form might have been as long as thirty feet from head to tail and weighed up to six thousand pounds. Crested duckbills are known as lambeosaurs, to differentiate them from such noncrested duckbills as *Maiasaura,* collectively referred to as hadrosaurs, so I named the bone bed Lambeosite. The end of the season was then fast approaching and in the little time that remained the crew could remove few bones from the two new sites. All the same, I drove back to the Museum of the Rockies certain where I would be digging the following year and more optimistic about the likely results than I'd felt about any excavation I'd started since leaving the Willow Creek anticline.

Anticipation was high when we returned to the Blackfeet Indian Reservation on June 20, 1986. There were sixteen of us—Bob, me, and fourteen students and volunteers, all itching to see what the teases of the previous season would lead to. We set up camp on the edge of the Milk River immediately north of Landslide Butte. Half of the crew then reopened Dino Ridge Quarry while the other half reopened Lambeosite, removing the dirt cover we always spread across our excavations as protection against the elements. At Dino Ridge Quarry it didn't take long to realize that we had unearthed an unusually rich bone bed. Before us lay an astounding concentration of fossils—vertebrae, ribs, arm and leg bones, teeth, skull fragments, as many as forty per square meter in some places, all mixed together haphazardly. With the exception of a few bones belonging to an unidentified juvenile duck-billed dinosaur and numerous tyrannosaur teeth, Dino Ridge Quarry was monospecific—that is, every bone came from the same kind of dinosaur, the new ceratopsian. An entire einiosaur herd, numbering in the dozens and perhaps more, had died in the same place at the same time.

What had killed them? Years ago, the presence of tyrannosaur teeth would have led some paleontologists to conclude that the herd of horned dinosaurs had been attacked by the relentlessly predaceous theropod of popular imagination. But as I explained in chapter 1, the anatomical evidence overwhelmingly suggests that *Tyrannosaurus* was a scavenger that possessed an exceptionally keen sense of smell and jaws well suited for tearing flesh but none of the other qualities required to chase down, subdue, and kill prey, especially a herd of two-and-a-half-ton behemoths bearing nose horns and shield spikes as long as a man's arm. No, what the teeth at Dino Ridge Quarry told me is that a number of very lucky tyrannosaurs had fed on the carcasses of a number of einiosaurs whose luck had run out.

The manner in which their luck ran out was later determined by Raymond Rogers, a graduate student in geology at the University of Montana, in a clever bit of reasoning that involved reconstructing the climatic and environmental conditions in which the einiosaurs lived and comparing them with similar conditions today. After analyzing the placement of the bones and studying the sediments and caliche layers in the area, Ray concluded that Dino Ridge Quarry represents a water-hole environment, a place where the herbivorous horned dinosaurs had taken refuge during an extended dry period along the upland coastal plain. As the drought worsened, vegetation dwindled further. Streams and shallow lakes dried up. More animals converged on the water hole, not only to quench their thirst but to forage on the plants surviving there. As those sources of food in turn became exhausted, the dinosaurs, now severely weakened, died of opportunistic disease, starvation, maybe even sunstroke.

Also undernourished because of the drought, various carnivores, including tyrannosaurs, found their way to the water hole to scavenge the many dead and dying animals. A large percentage of the fossil bones at Dino Ridge Quarry are scratched and broken, as if they had been trampled, adding credence to this idea. A virtually identical scenario can be observed during droughts in East Africa today: among large groups of elephants, for instance, that at first are attracted to a water hole because it offers both food and water, then linger too long, eating everything in sight and starving to death rather than leaving to find more food and thereby risking dying of thirst. Lambeosite resembled Dino Ridge Quarry in many important ways. It also was monospecific, containing only the bones of the crested duckbill, a new *Hypacrosaurus,* with the exception, once again, of an abundance of *Tyrannosaurus* teeth. And as was the case at Dino Ridge Quarry, the bones had been moved by water, but not very far, before they were buried. Finally, the overall sedimentary environment was very similar. Taken together, the evidence at Lambeosite suggests that there, too, a drought-related mass death had occurred.

As the two crews excavated Dino Ridge Quarry and Lambeosite, I surveyed outcrops throughout the badlands. Within two days I had discovered an ankylosaur skull and skeleton and the leg and foot of an *Ornithomimid,* an ostrichlike theropod unique to the late Cretaceous that stood about eight feet tall. In addition, halfway up the side of a small bluff I came across a narrow, lens-shaped deposit of mudstone that contained the remains of another duck-billed dinosaur. Judging from the skull parts at Westside Quarry (the name I gave this bone bed), which turned out to be monospecific as well, the animal appeared to be a previously unknown species of

Removing a clutch of eggs from the Landslide
Butte area. Ricky Reagan, whose land we were
camped on, is on horseback.

Prosaurolophus—a hadrosaur that grew to twenty-five feet in length
and weighed a couple of tons—also found only in late Cretaceous
sediments. Everywhere I walked, it seemed, I also found dinosaur
eggshells, countless small, black shards resting on the ground and jutting from outcrops. The specimens we recovered by removing hundreds of pounds of overburden—soil and rock lying on top of fossil
layers—at the Willow Creek anticline we could collect from a single
shovelful of dirt in some parts of the Landslide Butte badlands.

By that time, David Weishampel, a dinosaur paleontologist from
the Johns Hopkins School of Medicine, his family, and crew were

camping with us along the banks of the Milk River. I had invited David to explore a nearby section of the St. Mary River Formation, which represented upland coastal plain sediments deposited as the seaway receded for the last time during the late Cretaceous. While passing through the area in the early 1900s, Barnum Brown had collected a primitive horned dinosaur related to the Mongolian *Protoceratops*, which he had named *Montanaceratops*. David not only found a *Montanaceratops*, just a hundred yards from Brown's excavation, but also dinosaur egg and embryonic remains, the first ever from the St. Mary River Formation and another indication that green mudstones containing layers of caliche are excellent rocks in which to find such specimens. With David's discovery, and what my crews were uncovering in the uppermost layers of the Two Medicine Formation, I had ample reason to believe that I had come to the right place to test my ideas about parental care among dinosaurs. What's more, Bob and I had witnessed an omen—actually, two of them.

It may surprise you to hear a scientist talking about omens, but those of us who venture beyond the deliberately simplified, highly controlled conditions of the laboratory quickly learn that we are at the mercy of forces and events that lie outside our control—even, at times, beyond the reach of scientific understanding. Somewhat like farming, sailing, or any occupation that brings one into intimate and sustained contact with the elements, scientific fieldwork tends to instill in its practitioners a great respect for the vagaries and complexities of the natural world. Our fortunes, after all, are tied closely to such unpredictable phenomena as the weather. Like farmers and sailors, then, we invent rituals to manage the many uncertainties in our circumstances. This is not to say that when it rains I fall to my knees in the mud and pray that it will stop. But I have to admit that when the downpour does finally cease, I can't help but feel as if a small blessing has been bestowed upon me and my crews. And like farmers and sailors we sometimes indulge in the less-than-precise but nonetheless entertaining art of identifying signs of good fortune, of reading portents of things to come. For Bob and me, that had always meant keeping an eye out for blue herons and pelicans.

Over the years we had encountered so many herons and pelicans in the vicinity of our excavations that we came to expect to see them on a pretty regular basis. If we didn't, if what seemed like an unusually long time passed without our having spotted at least one, especially if our fieldwork wasn't yielding much in the way of interesting fossils, Bob might joke, "We shoulda known better, Jack, we shoulda known." He was especially fond of the pelican. One evening early in the 1986 season—I can't recall exactly when or even how the subject came up—he said that if he died sooner rather than later, his spirit would always remain close by, watching us through the eyes of a big white pelican. "And if I do die," Bob added emphatically, "give me a wake. None of that somber stuff." I remember thinking at the time that were one of us to go early, it would most likely be me. Bob just seemed unstoppable. As a young man he'd survived Hodgkin's disease, one of the first in the country to do that. His passion for fossil collecting was insatiable. He wouldn't think twice about walking twenty miles under a blistering sun or jackhammering limestone for twenty days straight. In camp Bob was a dynamo. Teacher, father figure, keeper of fieldwork stories and traditions, he initiated new crew members into the ways of paleontological excavation as well as into the secrets of living comfortably outdoors. Invariably, Bob was the last one to go to sleep at night and the first to wake in the morning, busily preparing supplies and equipment, filling water containers, getting everyone ready for another long day of digging.

So it was no surprise when early one morning in July, while the rest of the crew slept, Bob poked his head through the opening to my tepee and said, "Hey, Jack, you've got to see this." Living in tepees, by the way, was one of our fieldwork traditions. The structures are portable, fairly easy to set up, and you can cook in them if you have to. Most important, they are superior to all but a few tents in being able to withstand the winds that invariably whip across the high plains of central and eastern Montana. Our 1986 Landslide Butte camp consisted of several tepees pitched on a rise just above the river. Below, close to the bank, we had set up our kitchen tent, where everyone took meals together. And that is where Bob directed my attention when I stepped, a bit groggy, from the shadow world of my tepee into the hazy light of an uncomfortably warm morning.

"Look," he said. "There it is." Down in our kitchen area a great blue heron, evidently having come from the river, was tiptoeing around the tent as if trying not to wake anyone. Bob and I watched the bird for a while. Almost four feet tall, with very long legs and a long neck, the elegant wader seemed unfazed by our presence. In time, though, other members of the crew got up and moved about the camp, making noise. The heron took a couple of quick steps toward the river, unfolded its kite-size wings, and waved them up and down. Within seconds it was a hundred yards upstream, a swiftly receding blue-gray blur.

As if that weren't encouraging enough, about an hour later, during breakfast, a flock of white pelicans soared across the camp, low enough that I could hear the collective whoosh of their wings. Like most everyone, I enjoy the songs of birds, but there is something especially compelling about the sounds of wings in flight, something that tugs at the heart. In the moments when a large-winged bird passes close by one hears, I think, the coming and going of life itself, its lightninglike transience—the sudden rustling of the air, fast-moving shapes glimpsed out of the corner of an eye, followed by utter silence. As quickly as the white pelicans came into view, it seemed, they vanished over the horizon. "Gonna be a good day," Bob pronounced.

And it was, because that was the day I took a close look at Gilmore's datum layer, discovering that the fragments he had thought were the remains of freshwater clams were in fact dinosaur eggshells. Gilmore had chosen that layer of green mudstone, remember, not only because it was easily recognizable—by the fragments it contained—but because it stretched across most of the badlands north of Landslide Butte, providing a universal benchmark against which he could measure the distance and thus age of strata above and below it. In other words, when he misidentified the

The *Hypacrosaurus* nesting horizon at Landslide Butte. *X*s show the nesting horizon; the arrow points to the top of the Two Medicine Formation. (*Pat Leiggi, reproduced courtesy of the Museum of the Rockies.*)

eggshells he also missed the opportunity to see what so extensive a deposit really meant—that the fragments represent an immense nesting horizon, in fact, at a mile long and at least a quarter-mile wide, the largest dinosaur rookery in the Western Hemisphere. Averaging about two feet thick, this broad layer of mudstone actually contains numerous nesting colonies stacked on top of each other and belonging to the *Hypacrosaurus* we first saw at Lambeosite. The horizon contains thousands of nests, hundreds of embryonic and nestling hypacrosaur bones, and millions upon millions of eggshell fragments mixed together with whole eggs, squashed eggs, and egg clutches.

Apparently a herd of hypacrosaurs, consisting of at least a thousand adults, had gathered repeatedly in the Landslide Butte area to bear their young. Imagine standing on a hill and taking in that scene: twelve-foot-tall, thirty-foot-long crested duck-billed dinosaurs milling about as far as the eye can see. Now imagine this: heavy gray clouds approach from the south, originating somewhere within the Rockies. As they draw closer the sky darkens. It starts to rain, but what falls is not water. It is ash, a cataract of blinding soot and red-hot debris called airfall tuff. As has happened many times during the Cretaceous period, the volcanoes now known as the Adell Mountains, located in west-central Montana, are erupting, once again causing havoc among the dinosaurs that live on the upland coastal plains. And the eruptions will persist for days, perhaps longer.

The adult hypacrosaurs have no choice but to flee the rookery. Using their crests, which are hollow and of a piece with the nasal passage, the dominant males bugle a low-pitched alarm. Instinctively the herd responds, one by one abandoning their nests and moving in one gigantic formation to the east, or the north, away from the deadly fallout. The sound the heavy creatures make as they lumber across the land is thunderous. Soon the herd is engulfed in a cloud of dust. Ash continues to rain down. It seems as if the world is coming to an end. And it is, but only for the unborn and newly hatched dinosaurs left behind. The ash smothers the embryos and nestlings, burying eggs, nests, everything—the entire rookery. Today the rookery lies beneath a layer of rock called bentonite, a sediment formed from volcanic fallout. That's how we know what happened to the hypacrosaurs of Landslide Butte.

With the discovery of the *Hypacrosaurus* rookery, the upper layers of the Two Medicine Formation exceeded my hopes for the 1986 season. But our good fortune didn't end there. Toward the end of August, Carrie Ancell, a preparator who had spent two seasons at the Willow Creek anticline, and I discovered a second assembly of horned dinosaurs identical to those at Dino Ridge Quarry. Located about a mile from the first site, Canyon Bone Bed, as we called the new fossil horizon, comprised the same dark brown, sandy siltstone, arranged in the same depositional patterns, containing the same carbonized plant debris and remains of freshwater invertebrates, which led Ray Rogers to conclude that it, too, represented a shallow pond where a large group of ceratopsians had gathered during a drought, then died en masse. What was surprising about the new site were the skulls we found. At Dino Ridge Quarry we uncovered only cranial fragments, enough to suggest that they belonged to a new ceratopsian but insufficient to reconstruct the head in detail, enabling us to identify with confidence all the critical differences. Canyon Bone Bed, by contrast, yielded two nearly complete skulls—and one very intriguing puzzle.

The nose horns on most ceratopsians are quite similar. Those of *Styracosaurus* and *Centrosaurus,* for example, are long and straight, pointing upward, at a right angle to their snouts, and the same is true of *Triceratops*. But the nose horn on *Einiosaurus,* the first skull from Canyon Bone Bed, is short and hooked, curved so far forward that it points downward, toward the ground. What's more, instead of the six spikes fanning out from the neck shield, as is the case with *Styracosaurus,* the skull has a pair of long spikes that extend backward from its shield. A third distinguishing feature is a small, rounded nub over each eye socket. The nose horn of the second skull, recovered from a lower and therefore older level than Canyon Bone Bed, differs from that of *Einiosaurus* in that the tip curves ever so slightly backward; otherwise it is long and straight and pitched at a right angle to the snout, very similar to *Styracosaurus*. Yet, like the first skull, it possesses none of the other identifying cranial features of *Styracosaurus*. In particular, its neck shield also sports two long horns directed backward, and above the eyes lie bumplike protrusions.

Both skulls are moderate in size—about four feet long—and heavy, particularly when we encased them in plaster jackets, after which they weighed as much as a thousand pounds each, and that posed a problem. We remove most dinosaur fossils from our excavations using either a medical field stretcher or a device conceived and built by Bob, ever the restless gadgeteer. Nicknamed the Dino Wheel, it resembles a stretcher but is made of metal and is supported underneath by a motorcycle wheel riding on a short axle. Much refined over the years, our most recent model can still carry

Bob Makela with one of his field inventions, the Dino Wheel, which can be used by two people to get heavy loads in and out of areas having no roads. (*Pat Leiggi, reproduced courtesy of the Museum of the Rockies.*)

only about six hundred pounds of fossils. Making matters more difficult, the ceratopsian skulls were located on the face of a fairly steep cliff.

Solution? Call in the cavalry. Well, a modern-day version of the cavalry: the United States Army National Guard. A Huey gunship flew up from Helena, about two hundred miles away, to airlift the dinosaur skulls of Canyon Bone Bed into waiting trucks. For the guard pilots the exercise was an opportunity to practice picking up objects in rough terrain. For me, it was a source of amusement and wonder. The amusing part was watching giant blobs of plaster suspended from steel cables floating over the badlands of Landslide Butte. The wonder was aroused by the mysteries now hidden inside the plaster. That curved horn, especially. It was telling me something. But what exactly?

5

THROUGH THE EYES OF A PELICAN

A new duck-billed dinosaur, a new crested duck-billed dinosaur, two new horned dinosaurs, four bone beds containing the remains of juvenile and adult dinosaurs, and the largest dinosaur rookery this side of the Pacific Ocean. After collecting an array of specimens from the upper layers of the Two Medicine Formation during the 1985 and 1986 seasons, I decided again to explore its lower, older layers, specifically, those exposed in the southeast corner of the Blackfeet Indian Reservation, along the Two Medicine River, the formation's namesake. Located about sixty miles from Landslide Butte, the area contains sediments representing the entire formation, from bottom to top, some 12 million years of terrestrial deposition on the upland coastal plains. The reason so much of the geological record from the late Cretaceous is accessible there is that the Two Medicine Formation does not lie flat but tilts slightly, the uplifted side facing away from the mountains. During the 1987 season I planned to set up a camp near the formation's midsection, almost due south of Cut Bank, so that if we walked east, or downstream, we would encounter older strata, and west, or upstream, younger strata.

Possessing an uncanny sense of direction, that itinerant bonehound Barnum Brown had passed this way, too, in 1917, unearthing an ornithopod that resembles the late Cretaceous duck-billed dinosaur *Gryposaurus,* except that its teeth are flatter and broader, like those of *Iguanodon,* an ornithopod known only from the early Cretaceous. Brown's dinosaur possesses other primitive features as well, including unusually large forelimbs, all of which, taken together, indicate that it might have been an ancestor of some of the duck-billed dinosaurs. That, at least, was what I was led to believe upon examining the skeleton at the American Museum of Natural History in the late 1970s, and it was one of many ideas I hoped to test when we arrived at the Two Medicine River in early June. After reading Brown's hastily written field notes (he often scribbled in the margins of newspapers or scraps torn from paper bags), I was sure I could place my crew within a few miles of his original excavation, but that's all I could be sure of. Any reckoning more precise would be a matter of luck. And as sometimes happens, luck awaited us, in the person of Tom Harwood, gentleman, rancher, and, I've been told, consummate fiddle player. I approached Tom for permission to camp and search for dinosaurs on his property, which encompassed a promising stretch of badlands near the river.

"Young man," he said, "You can look for bones anywhere you care to, but I can't put you up in the spot where I usually put fellas like yourself. That washed away in the '64 flood."

The '64 flood? What did he mean by that? Did he remember where the "fellas" came from, their line of work?

"New York City," Tom replied. "A professor. Had a bunch of college boys with him."

I was dumbfounded. The only other paleontologist who had collected in the region was Barnum Brown. But was that possible? Had Tom met Brown?

"I guess it's been a few years," he said, chuckling to himself, then going on to explain that in 1917 he was seventeen years old. When Brown arrived at the family ranch, young Tom led him to a spot where a fossil jaw with large, flat teeth was located. Brown paid Tom a hundred dollars for his services as a guide and for the specimen. And it was that very specimen that I had studied at the American Museum. Small wonder that paleontologists are a little superstitious.

While Carrie Ancell and the first crew established a camp on Tom Harwood's land I sent a second crew, under Bob's leadership, back to Landslide Butte. Nineteen-eighty-seven brought a considerable expansion in the scope of my research, one made possible by an event that occurred the previous July. In Cut Bank to do our laundry, Bob, Dave Weishampel, and I had stopped at the local tavern for a couple of beers. While there I received a phone message from Mick Hager, at the museum, saying that I should call such-and-such number in Chicago and ask for a fellow named Hope. I figured Hope was a reporter who wanted a story about our fieldwork, so I didn't bother to return his call until I finished my laundry, when I learned that he was instead a representative of the MacArthur Foundation. "You've won a fellowship," Hope declared, with more drama than I could appreciate at the time, adding that I would receive a total of $204,000 over the next five years. Having never heard of the MacArthur Fellowship program and finding it more than a little suspicious that a man named Hope was offering me an enormous amount of money that I was free to spend any way I saw fit, I dismissed the episode as a practical joke.

The first check arrived in early August. I bought a new four-by-four pickup truck to replace the oil-burning, gear-slipping rattletrap I'd been driving all over Montana's back roads in recent years. In short order I also purchased much-needed equipment for the laboratory, including some of the most sophisticated computer gear available. I hired additional staff to oversee the cleaning and restoration of specimens brought back from the field and a part-time artist to provide illustrations, and eventually took on eighteen graduate students. To this day, I don't know how my name came to be included on the list of candidates the MacArthur Foundation considered for fellowships in 1986 or why exactly I was selected, but I'm ever grateful, because it gave me the wherewithal to seed one of the biggest dinosaur research programs in the world.

Buoyed by our success at Landslide Butte, the entirely unexpected MacArthur Fellowship, and my chance meeting with the only man alive who knew exactly where Brown excavated his primitive duck-billed dinosaur, I was in a highly optimistic frame of mind at the outset of the 1987 season. I should've known better. Fortune always goes the way of the wind, goes every which way at once. At Landslide Butte, Bob and his crew had reopened the two new horned dinosaur sites, Dino Ridge Quarry and Canyon Bone Bed, along with Westside Quarry, where we had found the new species of *Prosaurolophus*. Near the hypacrosaur rookery, which we called Datum Ash Layer, after the volcanic fallout that buried it, they soon found an incomplete adult skeleton and a nest with intact eggs inside. Then, on June 26, across the drainage from Westside Quarry, Bob discovered the partial skull of a *Pachycephalosaurus*, a very late Cretaceous dinosaur that, despite its superficial resemblance to the ornithopods, was most closely related to the ceratopsians. *Pachycephalosaurus* walked on two legs, stood about eight feet tall, and was fifteen feet long. But the dinosaur's defining feature is its skull—a very thick, high-domed braincase, from the back of which protrudes a row of bony nodules, attached to a short snout bearing small spikes.

Pachycephalosaurus fossils are extremely rare. Indeed, the dinosaur is known only from one complete skull, found in Montana in 1940, and a few dome fragments. Bob, understandably excited by his discovery and, as always, eager to share his enthusiasm with others, decided to drive down to the Two Medicine River camp to show the crew there the unusual specimen. I was back in Bozeman at the time. Long after midnight the phone rang. It was Gail McCrimmon, a volunteer from Alberta. "Something's happened to

Bob," she said, her voice oddly hollow, faltering. I held my breath. Gail explained that on the return trip to Landslide Butte earlier that evening Bob had stopped in Cut Bank to get gas, water, and supplies, then continued northward. At about 10 P.M., evidently having fallen asleep, Bob veered off the highway along a gradual turn just outside town. His truck rolled several times and the gas tank burst into flames. There was nothing anyone could do to help. A local rancher drove out to the Landslide Butte camp to tell the crew and to ask them to come into town to identify the body. Bob, my closest friend of more than twenty years, was gone. In a lightninglike instant, he had left us.

As Bob had requested, we marked his death by celebrating life, his and ours. On July 1, we held a wake for him at his house in Rudyard, the small town a hundred miles east of Cut Bank where he had taught high school science since graduating from the University of Montana. But it wasn't easy steering clear of the "somber stuff" that he had always considered a wasteful distraction. Besides inspiring a great deal of respect and affection, Bob had been one of the engines that made our summer expeditions go. He had been an ever-present, perpetually confident and confidence-boosting force. With his passing, we were in danger of losing momentum, of losing the will to continue the season. Those of us who had worked with Bob longest, however, knew that nothing would have disappointed him more than that. We knew that we would best honor his memory by doing what he most liked to do—collecting more dinosaur fossils. In the face of hard times Bob usually resorted to hard work, so that was our approach as well. I had already selected Pat Leiggi, my chief preparator, to take Bob's place at Landslide Butte, and he and his crew, eager to find something to take their minds off the loss, didn't wait for the wake to go back to work. Carrie, the other crew, and I had also resumed exploring and collecting along the Two Medicine River, always on the lookout for great blue herons and, especially, pelicans. Imagining that Bob was somewhere nearby, standing watch over the digs, made our newly diminished world easier to bear.

Fortunately, much occurred in the days immediately following Bob's death, keeping both camps very busy. Volunteer Sid Hoffsteder found yet another kind of ceratopsian skull near Canyon Bone Bed. Recovered from a higher, more recent layer of the section than the previous two, it possessed a neck shield with two long spikes thrust backward, like the hook-horned skull, but no nose spike. Instead of a knob in the vicinity of the eyebrows, the third skull had deep, rough gouges. Most unusual, the upper surface of the snout was ridged and gnarled and in all probability had been covered by a horny sheath, or boss, when the animal was alive, just as the horns of sheep and antelope are today. Only one other known horned dinosaur has a nasal boss—*Pachyrhinosaurus,* a late Cretaceous herbivore that grew to about thirty feet in length and, like most other horned ceratopsians, is found only in North America, along the Rocky Mountain Front. With the discovery of a third horned dinosaur, later named *Achelosaurus horneri,* in the upper layers of the Two Medicine Formation, a pattern appeared to be emerging, one, however, whose significance I was only beginning to grasp. When the three Canyon Bone Bed specimens are arranged according to age, the changes to the nasal horn and shield spikes seem to represent incremental developments among related species, as if the three animals belong to the same lineage. What's more, the intervening steps are visible at a level of detail never before observed among dinosaurs.

No sooner had I begun puzzling over that notion when we made a parallel discovery in the lower layers of the Two Medicine Formation, confirming my hunch about Brown's 1917 find. Earlier in the season, only a few dozen feet away from the spot where Tom Harwood had come across the specimen he sold to Brown seventy years earlier, Wendy Sloboda, an amateur collector from Alberta, had found an entire bone bed, now called Hillside Quarry. By the end of June the crew had recovered enough specimens from the site that we could be sure they belonged to the same dinosaur as Brown's, which I eventually named *Gryposaurus latidens,* or *Gryposaurus* with wide teeth. As I'd done with the horned dinosaurs of Canyon

Bone Bed, I arranged *Iguanodon, Gryposaurus latidens,* and other, more recent duck-billed dinosaurs according to age and once again a pattern became evident, especially as regards the teeth. The early Cretaceous *Iguanodon*'s diamond-shaped teeth stand side by side, two rows deep in the dental battery. The similarly shaped teeth of late Cretaceous duck-billed dinosaurs, by contrast, are narrower from front to back than *Iguanodon*'s and, most important, rest within batteries of four or five rows, producing a larger eating surface. *Gryposaurus latidens* falls midway between the two, with teeth that are narrower than *Iguanodon*'s but an eating surface that is smaller than that of the average duck-billed dinosaur. Coming from the bottom of the Two Medicine Formation, *Gryposaurus latidens* is of an intermediate age as well, suggesting that the changes to the teeth of the three groups of ornithopods are part of the same progression, among kin, from one generation to another.

As I say, back in 1987 it was unclear what these apparent patterns meant, or how much attention they deserved. We'd gone to the Blackfeet Indian Reservation, after all, to search for further evidence of social behavior among dinosaurs, especially with respect to reproduction and parental care, meaning, of course, more eggs, nests, and babies. Yet in pursuing that interest—specifically, by locating and excavating large monospecific bone beds—we came across all kinds of tangential but nonetheless fascinating information, which would have been foolish to ignore. In a word, our research was evolving, driven by what we happened to dig up no less than what we hoped to dig up. But that is the nature of paleontological fieldwork. It is an opportunistic endeavor, a mixture of design and improvisation, in which being prepared means being prepared to change course when one unexpectedly runs into a dead end or, as in this instance, when circumstances suggest another promising line of inquiry. Though neither Canyon Bone Bed nor Hillside Quarry contained evidence relevant to our ideas about parental behavior, they offered something every bit as valuable—the first, tantalizing clues to a new understanding of dinosaur evolution.

*   *   *

If additional clues lay ahead, however, they would be unearthed outside the Landslide Butte badlands, where, by the end of the 1987 season, we had collected most of the accessible fossils—accessible, that is, without the assistance of large earth-moving equipment, which surely would have disturbed Ricky Reagan's peaceful existence, to say nothing of the impact on his land. At the end of August we loaded a grain truck with our cache of dinosaur bones, including Bob's *Pachycephalosaurus,* the third horned dinosaur skull from Canyon Bone Bed, and the *Gryposaurus latidens* remains from Hillside Quarry, and drove back to Bozeman. In addition to the usual winter activities at the museum—cleaning and preparing specimens, studying them with computerized tomography (CT) scans and high-powered microscopes, writing professional papers describing our discoveries—I mounted an effort to hire a combined cook and camp manager. In years past, everyone pitched in and helped with general chores while the job of preparing meals rotated from individual to individual, but that wouldn't do during the 1988 season, when I expected to field my largest crew ever near the Two Medicine River. Mel Jones, who had cooked for pack trains of hunters in the mountains of western Montana, took the job.

In the second week of May we set up camp on the south bank of the river about a quarter mile from Lewis and Clark's so-called Fight Site, where, in 1805, the exploration party shot and killed a Blackfeet warrior, the first and only Indian to die at their hands. With eleven Blackfeet and Sioux tepees, several tents, trailers, and trucks, and a crew that varied between thirty and forty people, our encampment was an impressive sight, especially when viewed from Hillside Quarry, located nearby on a high cliff. Visible, too, from the quarry was a long stretch of the Two Medicine River, as well as much of the wildlife that made the riverine environment their home, including pelicans, which were present in the area when we arrived and remained throughout the summer. Although a year had past since Bob died, starting a season without him, without his energy, was disorienting. He and I had collected dinosaur fossils together since the early seventies, when all that we needed to mount an expe-

The 1988 camp along the Two Medicine River. (*Bruce Selyem, reproduced courtesy of the Museum of the Rockies.*)

dition was a couple of rock picks, a tank full of gas, and a cooler of beer, and the only payoff we sought was the thrill of discovering something new. What would he make of this expedition, of all the equipment, people, and ambition arrayed before me in the camp below? And how did it look, I wondered, to the large white pelican just then spiraling downward on angling wings, toward the river, landing with a splash alongside its fellows?

Among those to begin working immediately at the newly reopened Hillside Quarry was Ray Rogers, who recorded the strata at the site, as well as those in the surrounding area, just as he had done the previous summer at the monospecific bone beds near Landslide Butte. I'd assigned Ray the task of mapping the Two Medicine Formation, to enable us to compare and correlate the particular sediments in which we found dinosaur fossils. The geologists who had first surveyed the formation, determining, for example, that it is about two thousand feet thick along the Rocky Mountain Front, didn't identify and measure individual strata at the level of detail needed to determine their exact age and the specific types of environment they represent. Ray's job, then, was to increase the resolution, to bring into focus as many of the layers as possible, and there are hundreds of them. He measured their thickness, analyzed their sedimentary composition and structure, identified their nondinosaur fossil contents, then recorded the positions of all of our excavations in terms of the vertical layers and with respect to each other. This is how we discovered, for example, that Hillside Quarry lies at the same level as Red Rocks, just south of the Willow Creek anticline, which means that the two sites are approximately the same age and, further, that the duck-billed dinosaur *Gryposaurus latidens* from Hillside Quarry and the *Protoceratops* from Red Rocks most likely lived at the same time, about 80 million years ago, or about four million years before the *Maiasaura* and *Troödon* we recovered from the anticline and six million years before the horned dinosaurs of Landslide Butte.

With Ray mapping the Two Medicine Formation and the crew excavating Hillside Quarry, Carrie, now my senior preparator, and I explored the cliffs bordering the Two Medicine River in the vicinity of the camp. We found a few isolated bones and teeth, but the best specimen of the search actually found us. It was Friday, June 3. We had just returned from Cut Bank when we spotted an unfamiliar man walking in the badlands below the cliffs. After watching him through binoculars for about a half hour, Carrie, Pat, and I hopped into my truck and drove to the top of the cliff to see if he had been given permission to enter reservation land and, more important, to find out what he was looking for in "our" backyard. I clambered down a cliff and made my way to the spot where he had last been walking. When I finally caught sight of him again, he was still a hundred yards away and standing on a ledge below me, not so far, though, that I didn't immediately notice the .357 Magnum pistol strapped to his hip, along with a backpack containing fossil digging tools. In the most nonaggressive voice I could muster, I yelled, "Hey, what're you doing out here?"

"What're *you* doing here?" he yelled back.

Ray Rogers (*right*) and April Lafferty measure the thickness and study the sediments of the Two Medicine Formation along the Two Medicine River. (*Bruce Selyem, reproduced courtesy of the Museum of the Rockies.*)

I told him that we were dinosaur hunters from Bozeman, adding that the landowner, Tom Harwood, had asked us to keep an eye out for trespassers.

"I'm just picking up pieces of bone," he replied.

I then climbed down to the ledge and introduced myself to Bob Harmon, a part-time carpenter from Cut Bank. Bob explained that he collected fossils to show to kids in the local schools. Judging from the way he talked, it seemed he knew more than a little about dinosaurs, where to find their remains, and how to remove them from the ground. Just then Pat and Carrie arrived with a six-pack of beer. The four of us sat down on a sandstone boulder and talked with Bob about the remainder of his summer. I asked him if he would like to work with us, collecting dinosaur fossils for the Museum of the Rockies. Bob, I would soon learn, is not one to display emotion in public but upon hearing my invitation he couldn't contain his excitement. I don't know which of us was more surprised by the outcome of our accidental encounter or, in retrospect, more pleased. I like to think that somewhere in the eroded hills along the Two Medicine River that day, a pelican was watching over the entire exchange, a bemused and approving glint in his eyes. Industrious, loyal, increasingly knowledgeable, Bob Harmon has worked for me ever since. When, recently, Pat took an administrative position at the museum, I made the former amateur collector from Cut Bank my chief preparator.

Shortly after Bob Harmon joined the crew we were forced to abandon our exploration of the older, lower layers of the Two Medicine Formation. We had spent a week crawling over the sedimentary beds east of camp, without finding anything of real significance, or anything of real significance that was readily recoverable. One bone bed, whose existence we deduced from fossils jutting from a single layer in a steep bluff, lay underneath sixty feet of rock, which is to say, out of reach. We also found a few sites that contained concentrations of eggshell but, as happened in the Red Rocks area, we uncovered no whole or even partial eggs, nor any baby bones. It soon became clear that Hillside Quarry represented the oldest rock in which we were going to be able to excavate useful specimens on the Two Medicine River and that if we hoped to find more dinosaurs we had better start looking at the younger beds in the badlands to the west. We had reached another of the dead ends that make the endless maze of paleontological fieldwork so unpredictable.

With the change in direction came a dramatic change in luck. While hiking through a rugged area about a mile and a half west of camp, I spied a brownish layer of mudstone from which hundreds of bones were weathering out. Almost all of the skeletal remains belonged to the same species of duck-billed dinosaur and, more exciting yet, almost all of them came from juveniles. I returned to camp to fetch part of the Hillside Quarry crew. Shortly after we began excavating we uncovered cranial bones of juvenile *Maiasaura*. Apparently, West Hadrosaur Bone Bed, my name for the site, lay at the same stratigraphic level as the Willow Creek anticline. Further exploration in the area turned up a few skeletal remains of *Orodromeus* as well. The water patterns in the sediments, the presence of freshwater snails, clams, and plant remains, along with the orientation and undamaged condition of the fossils, suggest that West Hadrosaur Bone Bed probably had been a small lake environment and that, like the three new species of ceratopsians at Dino Ridge Quarry and Canyon Bone Bed and the new species of *Prosaurolophus* at Westside Quarry, the *Maiasaura* there died of drought-related causes. Why exactly only juveniles are represented remains a puzzle. Perhaps they lived together, segregated from other age groups, but even if that were true in general it would have been improbable along the banks of a small lake or water hole, where entire herds of the same species would have gathered. A more likely explanation is that the younger, more vulnerable members of a starving *Maiasaura* herd succumbed to opportunistic disease, predators, and so on, while the adults survived. It's also conceivable, of course, that we were looking at a small piece of a much larger bone bed and that given more collecting we would have found adult remains.

After constructing the best possible picture of what occurred at West Hadrosaur Bone Bed I told the crew to proceed with the excavation while I surveyed the sedimentary rock farther west. I admit that I don't have the patience I had back in the days when Bob Makela and I would labor in the same quarry day after day, all day long, removing minute amounts of rock and dirt with chisels and whisk brooms. But the chief reason I'm so quick to leave a new dig and begin exploring again is that I'm interested less in the bones per se than in what they reveal about large-scale trends. While not entirely consistent, for example, the combined ecological evidence from Landslide Butte and Two Medicine River strongly suggested that the horned and duck-billed dinosaurs living on the upland coastal plains during the last expansion of the Western Interior Cretaceous Seaway were especially susceptible to changes in their environments. Instead of a single snapshot of mass death at one water hole, in other words, we had over the course of four field-work seasons assembled a story, or a part of one, in which herds of dinosaurs all along the upland plains were not only being crowded into a smaller and smaller habitat—that ever-rising sea encroaching from the east—but also were periodically subjected to severe reduc-tions in food sources. According to my reading of the upper layers of the Two Medicine Formation, the period from 75.4 to 74 million years ago was a very stressful time for dinosaur populations along the Rocky Mountain Front. And having understood that much, I wanted to read on, to find out more about the lives of late Cretaceous dinosaurs. So, while the crew removed the remaining dust from one page I turned to the next one, walked over the next hill, where the sedimentary beds were younger still.

And those beds are located on land leased by the Carroll fam-ily—Vernon and his parents, Louis and Rose. I met Vernon first. A vigorous, outgoing man, he jumped down from his tractor, shook my hand firmly, and asked why it had taken so long for me to find my way to his property. Evidently, as had happened at the Willow Creek anticline, news of our activities at Landslide Butte and the Two Medicine River was making the rounds among the residents of the Blackfeet Indian Reservation. Vernon said that he had been looking forward to meeting me and before I could get the now well-rehearsed words out of my mouth he gave me permission to explore and excavate on all of the Carroll land. He added that he had never thought much about dinosaurs but was very interested in the mili-tary forts and trading posts that had been erected along the Two Medicine River during the nineteenth century. I promised to keep an eye out for anything of historic or archaeological value but I saw no such artifacts while working in the region, perhaps because my search image is set to a much earlier era.

About six miles upstream from camp, however, at the mouth of Badger Creek, near the western boundary of the Carroll property, I came across outcrops that looked very much like the sediments in which we had been finding fossils all summer long. My son Jason, then sixteen years old, had joined the Two Medicine River expedition for a few weeks, and he and I returned together to investigate the area more closely. Still hunting for evidence relevant to our ideas about parental care, I told Jason to look very carefully for eggs and babies; having spent several summers at the Willow Creek anticline, he knew exactly what I had in mind. Then we separated. About an hour later I discovered a nearly complete skeleton of an embryonic *Hypacrosaurus*. Most of the surrounding mudstone had weathered away, making it fairly easy to pry loose the fragile specimen, a whole dinosaur only eighteen inches long. Just as I began ever so gently removing the tiny bones from the rock, Jason hollered at me from two hundred yards down the enbankment. He had found part of an adult dinosaur, he said. I shouted back that he should let it be and go find a baby. Please. I wanted to see babies. But Jason insisted. "Dad, you better come see this," he said. "I think you're really gonna like it."

I was in no mood to leave my precious infant treasure for what in all likelihood would turn out to be a large and cumbersome leg bone whose owner I wouldn't be able to identify anyway. But remembering the mosasaur Bob and I had reluctantly gone to see at Four Horns Lake, to say nothing of the many times I've been wrong in these matters, I walked to the small hill where Jason was standing. And what I saw there I liked, really liked. Jason had found a large-scale version of what I had been placing in a plastic Ziploc storage bag. Poking out of the side of the hill was the hind leg of an adult *Hypacrosaurus*. Scattered nearby were large vertebrae and portions of the dinosaur's front legs. The skeleton appeared to be complete. There was a problem, though: to remove many of the bones, including the skull, which was not visible, a great deal of overburden would first have to be cleared away. I brought the entire Hillside Quarry crew, under Pat's direction, out to Badger Creek to help with the first, most laborious steps of that process. Then Bob Harmon and I went for a stroll.

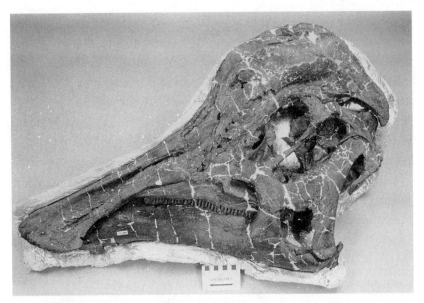

The skull of an adult *Hypacrosaurus stebingeri* found by Jason Horner in 1988. (*Bruce Selyem, reproduced courtesy of the Museum of the Rockies.*)

Well within shouting distance of Jason's Giant Site, my son's name for the new quarry, we made a very different kind of discovery—a ridge about one hundred yards long and twenty yards wide, both sides of which were paved with dinosaur fossils. Glittering in the afternoon sunlight were thousands of bone fragments belonging to at least several kinds of dinosaur, including, at first glance, hadrosaurs, lambeosaurs, and tyrannosaurs. And that much we determined from a brief visual survey during which Bob and I said little more than "Wow!" and "Can you believe it?" Every bone bed we had found before had been monospecific, exclusive, dominated by one type of dinosaur or another. This ridge, by contrast, which seemed to contain representatives of every animal that inhabited the upland coastal plains during the late Cretaceous, was a place where any creature, regardless of pedigree, could come to die. I'd never seen anything like it.

I called the crew over from Jason's Giant Site. No less astonished than Bob and I, they immediately fell into the game of trying to identify as many types of animals as possible. Since it was June 15, 1988, the day I turned forty-two, Bob suggested that we call the graveyard Jack's Birthday Site. Good idea, I thought, and a more satisfying birthday I couldn't have imagined, save for one thing: There was another face I would have enjoyed seeing on the ridge that day, another voice I would have enjoyed hearing utter once again, "Jack, it's time for a beer."

6

WALTZING TO
THE RHYTHM OF
THE WESTERN SEA

By mid-June 1988, after almost five weeks of digging, the crew at Hillside Quarry had recovered three skeletons of *Gryposaurus latidens*—an adult and two juveniles. But the amount of overburden had increased rapidly as they dug into the steep slope, trying to expose more of the narrow sedimentary layer in which the bones were entombed, and it now represented too great an obstacle to further excavation. I closed the quarry and transferred everyone to our new digs. From West Hadrosaur Bone Bed came dozens upon dozens of skeleton fragments, representing at least nine half-grown maiasaurs. An intact, undamaged hypacrosaur skull, two and a half feet long, was soon unearthed at Jason's Giant Site. And at Jack's Birthday Site the fossils were so plentiful that it took a crew three days to record and collect what lay on the surface alone. They then started to excavate in earnest the green-gray mudstone that forms the staggered backbone of the ridge.

Once again I set out to survey more of the region. Badger Creek joins the Two Medicine River at a spot where the river turns northward. The creek, then, runs from west to east, just as the river does farther downstream. Judging from its stratigraphic features and the dinosaur remains it contained, Jack's Birthday Site is located within the upper three hundred feet of the Two Medicine Formation, the same period of time represented by the sediments of Landslide Butte. Further evidence that the two are contemporaneous lay immediately west of the site—Bearpaw Shale marine deposits, marking the top of the formation and the end of dinosaur hunting along Badger Creek. Having discovered this, my first inclination was to explore badlands that parallel the creek drainage—that is, run from west to east, due north of Jack's Birthday Site. And I thought I found what I was looking for along the river about three miles upstream of the mouth of Badger Creek, on land leased by Truman Hall, an Indian who raises bucking horses and Brahma bulls for rodeos. During my first visit to the Hall ranch I turned up a few specimens, including baby bones, and several potential egg sites, but nothing that got me excited.

Accompanied by Carrie and Jason, I next visited an area near Blacktail Creek, about ten miles to the south and, ironically, only a few miles from Four Horns Lake, where three years earlier Bob Makela and I had excavated the mosasaur. I chose the spot because in his field notes Barnum Brown had described collecting there, briefly, and retracing his hurried footsteps had been a fruitful strategy ever since I arrived at the Two Medicine River. The move to Blacktail Creek would prove to be no exception. On June 23, our first day in the area, Jason and I discovered a large accumulation of baby hypacrosaur bones, jumbled together in what appeared to be a stream-deposited sandstone. In other words, the bones had washed in from elsewhere, though the absence of scratched or broken fossils indicated that they hadn't been carried over a large distance. Upon further exploration we discovered why they hadn't traveled far—the ancient stream had meandered alongside a hypacrosaur nesting ground. At Blacktail Creek North, our name for the site, Jason and I found several discrete collections of eggshell that almost certainly were the remains of nests. Then Carrie made an outstanding find—a clutch of eggs containing skeletal fragments of hypacrosaur embryos.

Roughly two hundred yards square, the Blacktail Creek North nesting ground is only about one-tenth the size of Datum Ash Layer, the rookery at Landslide Butte. Blacktail Creek North also differs from its northern counterpart in that the skeletal elements are concentrated in an area where the ancient stream abruptly widened, leaving behind a splayed deposit, whereas those at Landslide Butte are distributed randomly across the entire horizon. And at Landslide Butte, of course, the rookery lies beneath a layer of volcanic ash, an immense number of eggs and small nestlings having been buried alive. The cause of death at Blacktail Creek North is unknown, but given the preponderance of drought-induced die-offs recorded in the upper layers of the Two Medicine Formation, it wouldn't be surprising if that were the case here as well. Despite their differences, however, the two sites are similar in the ways that count most. For one thing, they are the same age. For another, they contain the remains of the same dinosaur, which Philip Currie, of the Royal Tyrrell Museum of Paleontology, and I eventually named *Hypacrosaurus stebingeri*, to honor the memory of Eugene Stebinger, the geologist who first described the Two Medicine Formation.

In the family drama we witnessed at the Willow Creek anticline, *Maiasaura* had held center stage, with *Orodromeus* and *Troödon* playing crucial supporting roles. But at both Landslide Butte and the Two Medicine River, in sediments laid down 2.7 million years after those of the anticline, *Hypacrosaurus stebingeri* stepped into the spotlight while *Maiasaura*, though still present and important, withdrew into the background. From that rainy August afternoon back in 1985 when I discovered the first bones of the new crested duck-billed dinosaur in a rain-soaked patch of ground I christened Lambeosite, we had encountered *Hypacrosaurus stebingeri* just about everywhere we looked in the top of the Two Medicine Formation—the Landslide Butte rookery, Jason's Giant Site, Jack's

Birthday Site, and now Blacktail Creek North. Not only had we found the new dinosaur in abundance, we had recovered complete skeletons representing the major stages of its life cycle—embryos, juveniles of several different ages, and adults. Rarely is so complete a sample ever assembled, never mind as quickly as this. And most exciting of all, we had found nesting colonies, two of them, in fact, which told us that *Hypacrosaurus*, like *Maiasaura, Orodromeus,* and *Troödon*, was a gregarious dinosaur, a social animal. It routinely gathered with its own kind to build nests and to lay its eggs. That now was indisputable.

But what about other aspects of parental behavior? Did hypacrosaur adults brood? Did hatchlings remain in the nest, their parents feeding and protecting them until they grew large enough and strong enough to fend for themselves? With respect to these questions, the available evidence, though copious, remained

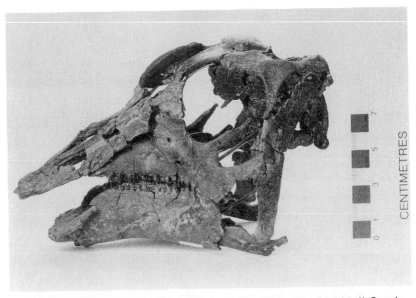

Skull of a juvenile *Hypacrosaurus stebingeri* from the Blacktail Creek nesting site. Note that the juvenile does not have the extensive nasal crest seen in the adult. (*Bruce Selyem, reproduced courtesy of the Museum of the Rockies.*)

ambivalent. The specimens present in our hypacrosaur excavations constituted the greatest volume of egg, nest, embryo, and juvenile material of any species of duck-billed dinosaur in the world. Yet we hadn't found a single hatched baby within a nest, as we had at the anticline, the breakthrough discovery that had convinced me that *Maiasaura* tended to the needs of its young following their birth. That in itself was puzzling. But, using the Null Hypothesis approach to prove my own ideas wrong, was it reason enough to abandon the parental care hypotheses altogether, at least insofar as *Hypacrosaurus* was concerned?

No, it wasn't. As I say, the evidence was ambivalent. Equally puzzling, and possibly contradictory, was the limited range of ages represented by the skeletons in the Landslide Butte rookery and at Blacktail Creek North. They consist exclusively of embryos, confined largely to nests, and juveniles up to about four feet in length, all found within the boundaries of the nesting ground. At neither location did we find larger juveniles, subadults, or adults. By the same token, Lambeosite and Jack's Birthday Site contain no hypacrosaur skeletons of nestling size; the smallest juveniles in those bone beds are considerably larger than the largest juveniles in the nesting grounds. Merged into a single, coherent picture, the two lines of evidence would seem to suggest that hypocrosaurs left their nests upon hatching but remained within the nesting ground until they reached a certain size, after which they departed the colony, joining herds comprised of subadults and adults. If they ever returned, it was only after they had reached breeding age themselves and instinctively found their way back to their native nesting grounds, a scenario that certainly is consistent with the overlapping colonies at Landslide Butte.

I had a hunch why only small juveniles were present in the colonies but I required additional information before I could be reasonably sure. Some of what I needed to know could come only from examining the bones of the embryos and nestlings much more closely, using the computer-aided imaging devices and microscopes back at the museum laboratory, a project that would have to wait till the winter, or perhaps the winter after that, depending on how quickly the specimens were cleaned and prepared. At the moment, a more pressing task presented itself—getting the specimens out of the ground. Although I'd assembled the largest field crew ever, all of them were occupied at our other sites, leaving only Carrie, Jason, and me to excavate the hypacrosaur nesting colony. There would be no more scouting trips for me in 1988. It took the three of us three full weeks to unearth the accessible fossils, representing at least eighteen juvenile dinosaurs, which we packaged in three five-hundred-pound plaster jackets. By that time, the last of the adult hypacrosaur bones had been removed from Jason's Giant Site and the excavation was complete. At the end of the month we closed down the rest of the quarries for the season and returned to Bozeman, but not before one last, haunting discovery.

Driven less by the expectation of finding something important than by a stubborn reluctance to quit looking, I returned to another site on Blacktail Creek on the final day of July, and there I located and collected a nearly complete *Troödon* skeleton, the first ever from North America. *Troödon* is a curious little theropod. From head to tail it was about six feet long. Rearing back on its hind legs—*Troödon* was bipedal—it stood no more than four feet high, if that. But like its distant cousin, *Deinonychus*, it was an aggressive carnivore, though not nearly as fierce or effective, probably preying on lizards, small mammals, and young dinosaurs such as the *Orodromeus* juveniles that lived at Egg Mountain, where *Troödon* teeth are plentiful. *Troödon* had a relatively large brain and, perhaps most interesting, large, forward-facing eyes, leading some paleontologists to believe that it possessed stereoscopic vision. Indeed, only a few years before I found the skeleton, Dale Russell, then with the Canadian National Museum of Nature, in Ottawa, speculated that had *Troödon* survived, it might have given rise eventually to a sapient, two-legged, upright creature that resembles human beings, one that could be sharing the planet with us today—or, more sobering, might have superceded us.

That, of course, is the sort of imaginative exercise that is great

Troödon formosus was a predatory dinosaur with good grasping arms and hands and a mouth full of coarsely serrated teeth.

fun to perform but whose results are impossible to verify. Yet as I gazed into the oversize eye sockets of the *Troödon's* birdlike skull, trying to imagine the long, slender snout receding and the low forehead growing wider, I found myself wondering what exactly the world would look like today if some of the events that shaped the natural history of dinosaurs had taken a slightly different turn. In my opinion, the extinction of *Maiasaura, Hypacrosaurus,* and their kin was neither as neat nor as unavoidable as the doomsday meteor theory would have us believe. Like all organisms, dinosaurs were subjected to a complex and dynamic array of forces, and though the forces are now largely understandable in and of themselves, their consequences, being dependent on untold numbers of contingencies, were not always or even mostly inevitable. Something as unimportant, seemingly, as a small shift in the climate—producing fewer droughts along the upland coastal plains, for example—could have profoundly altered the fortunes of certain dinosaur lineages in the late Cretaceous period. When I look into the holes where the *Troödon's* eyes once looked out at an earlier and very different world, I don't see a reflection of myself, as those who study primates often claim, but I do see a reflection of something more fundamental: my condition, our condition, the tenuous hold all living things have on existence. And but for the aimless grace of natural history, there lay the last member of my kind, buried in a mudstone graveyard, yet another evolutionary experiment come to an end.

The *Troödon* skull was not the only thought-provoking fossil among the new additions to our collection. With the approach of winter came our seasonal change in activity and orientation. Home from the hunt, back from the excitement of the chase, we would retire, somewhat fatter after ten weeks of Mel's fabulous meals, to the laboratory and for the next several months ponder our specimens, methodically studying their every aspect—their shapes and sizes, internal makeup and structure, how they fit together as full skeletons, how the animals the skeletons represent might have moved, their probable behavior, the roles they played in the ecosystems they inhabited. What did the fossils tell us about the lives of dinosaurs on the coastal plain during the late Cretaceous? That was the question behind everything we did during the winter of 1988–1989. While Pat, Carrie, Bob, and preparator Allison Gentry, assisted by volunteer Bea Taylor, began the laborious task of removing the plaster jackets and preparing the specimens, I reviewed the results of our four seasons on the Blackfeet Indian Reservation.

Overall, the most notable development since leaving Egg Mountain concerned the thrust of our research: It now ran on two equally important tracks, neither of which showed signs of ending soon. The first stretched all the way back to our earliest days at the Willow Creek anticline and carried through to our most recent excavation—the nesting horizon at Blacktail Creek North. In *Hypacrosaurus stebingeri* we had found a dinosaur whose behavior we could compare with that of *Maiasaura*. It wasn't a perfect match, to be sure, but the hypacrosaur material was so comprehensive and rich I felt confident that by the time we completed our study of it, teasing out all of the relevant implications, we would be able to say whether our original ideas about reproduction and the treatment of babies among duck-billed dinosaurs had been wrong.

The second research track started at Dino Ridge Quarry, in the Landslide Butte badlands, ran through Canyon Bone Bed and on to Hillside Quarry along the Two Medicine River. The three new horned dinosaurs and the new gryposaur gave me reason to believe that we might be able to answer questions that no one before had been able to address, much less resolve: What was the rate of evolution among dinosaurs? What were the mechanisms that drove their evolution? Our journey along this track had scarcely begun, however, and I was eager to move ahead, to see what lay around the next bend. Specifically, we needed more fossils, as well as a more thorough understanding of the large-scale environmental events that occurred on the coastal plains during the Cretaceous period, which is to say, we also needed to see more sedimentary rock.

So the following season, the summer of 1989, we went back to the Blackfeet Indian Reservation for what would turn out to be our last major field effort in the Two Medicine Formation. As part of his doctoral research, graduate student Scott Sampson had taken an interest in the horned dinosaurs of Landslide Butte. He wanted to investigate in particular what role the diverse headgear played among such highly social, plant-eating dinosaurs. Were the neck shields and spikes weapons of defense against predators? Or were they used primarily for sexual display and nonlethal sexual competition between rival males, as is the case among gregarious horned animals today? In the second week of June, Scott and a small crew reopened Canyon Bone Bed, while Pat Leiggi and another small crew assisted in the effort by reopening Dino Ridge Quarry. Meanwhile, graduate student David Varricchio, Allison Gentry, and a third crew established a camp on the Carroll Ranch along Badger Creek, then resumed the excavation of Jack's Birthday Site. Also working on a dissertation, Dave planned to spend the summer trying to determine the type of environment the site represented and what exactly happened to the animals assembled on that ridge. Jason, Carrie, and I pitched our tents on Birch Creek, which forms the southernmost border of the reservation. From there we could reach the Blacktail Creek region much easier than from the river, the approach we had used the previous two seasons. Our camps were spread from one end of the Blackfeet Nation to the other, yet all of us were either exploring or excavating within the top three hundred feet of the Two Medicine Formation.

During the next several weeks the exploring party—Jason, Carrie, and I—found a number of important specimens in the vicinity of Blacktail Creek: the skull and partial skeleton of an *Achelousaurus* and numerous areas containing *Hypacrosaurus* and *Troödon* eggs, embryos, and babies, along with several *Hypacrosaurus* and *Prosaurolophus* bone beds. Tiny islands of rock surrounded by vast sea-swells of grass, the new sites were not promising enough, however, to warrant further digging, at least not until we had finished excavating our other quarries, principally, Jack's Birthday Site, where Dave, Allison, and their crew were recovering the remains of almost every kind of dinosaur that had been found previously in the upper part of the Two Medicine Formation. Besides fossil concentrations of *Troödon, Hypacrosaurus stebengeri,* and *Prosaurolophus blackfeetensis,* the name I gave to the new duck-billed dinosaur first seen at Westside Quarry, near Landslide Butte, they found scattered skeletal elements representing, among others, ankylosaurs and *Daspletosaurus,* a theropod very similar to *Tyrannosaurus,* though not quite as large. Nondinosaur fossils included turtles, whole freshwater fish, frogs, lizards, birds, and pterosaurs (the flying reptiles that were close relatives of dinosaurs). In an area between Jack's Birthday Site and the Badger Creek camp, graduate student Vicki Clouse discovered an intact *Daspletosaurus* skull and a *Daspletosaurus* leg, complete from thighbone to claws. No other parts of the animal could be found. Bob Harmon and a small crew excavated the specimens during the final days of the 1989 season.

Also by that time, Dave had assembled a tentative but fascinating scenario for Jack's Birthday Site. As Ray Rogers had done at the Landslide Butte quarries, he measured and characterized the sediments in and near the ridge and conducted a taphonomic investigation of the fossils uncovered at the site, which is to say, he mapped the positions and conditions of the bones. Taphonomy,

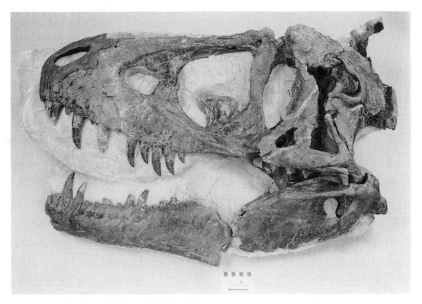

Skull of the tyrannosaur *Daspletosaurus* found by Vicki Clouse near the Two Medicine River. (*Bruce Selyem, reproduced courtesy of the Museum of the Rockies.*)

the study of what happens to the remains of plants and animals after they die, is a crucial tool in determining as well how the plants and animals died, even, in some cases, how they lived. The mere fact that all of the bones at Jack's Birthday Site are disassembled—there are no articulated skeletons—tells us that movement of some kind has taken place. Indeed, movement is so common among fossils that very rarely do we find intact, undisturbed skeletons, such as the hypacrosaur at Jason's Giant Site or the brachylophosaur we dug up outside Malta in 1996, which is why those excavations, though involving only one dinosaur, were well worth the effort. The challenge, of course, lies in determining the processes by which a skeleton has been disassembled, somewhat in the same way that a forensic pathologist goes about reconstructing a murder and its aftermath.

If a bone exhibits teeth indentations, for example, it could mean that it had been carried to its present site by a predator. At the very least, we can be sure that the animal to whom the bone belonged was either preyed upon or scavenged, activities that in themselves are sufficient to disturb a carcass and, thus, fragment and disperse the skeleton. As we saw in Ray's analysis of the horned dinosaur bone beds, certain kinds of nicks and breaks are highly indicative of trampling. Transportation by water produces characteristic abrasion patterns. Also distinctive are the effects of weathering, when bones lie on the surface for a long time before they are buried in sediment and fossilized. Any fossil whose outer surface is pristine, free of telltale weathering marks, was almost certainly buried rapidly, and on land that usually happens in a floodplain setting adjacent to a stream, water hole, or lake. At Jack's Birthday Site, Dave identified two general groups of bones, each with its own taphonomic history, or death story. The first and substantially smaller of the two represents a very wide range of animals whose remains are randomly dispersed. The condition, location, and orientation of the bones suggest that these accumulations occurred over a period of one hundred to one thousand years and resulted from everyday attrition among animals living at the site and the occasional arrival of skeletal elements washed in by any of several streams, some of which may have flowed only at certain times of the year.

The streams, Dave decided, converged on a small, shallow floodplain lake, and it is within the immediate vicinity of this lake, on a single horizon representing a period as short as one year or less, but no longer than ten years, that the second group of specimens is located. These consist of clusters of disarticulated but associated bones, that is, skeletons that have been disassembled but nonetheless remain segregated by species. Of the ten types of dinosaur recovered from Jack's Birthday Site, three are found in such segregated clusters—*Hypacrosaurus, Prosaurolophus,* and *Troödon.* (The *Troödon* remains, by the way, represent at least four individuals, the first time this dinosaur has been seen in the company of its own kind.) The death story for the second assemblage of fossils is more difficult to reconstruct. That the clustering of remains occurred during a very brief time—a single season possibly—suggests that the dinosaurs within each cluster died all at once and from the same cause, as had the ceratopsians at Dino Ridge Quarry and Canyon Bone Bed and the prosaurolophs at Westside Quarry. But the diversity of the dinosaurs in the clusters and their spatial distribution raise the possibility that a series of mortality events occurred, events that may or may not have been related.

In any case, drought almost certainly contributed in some measure to the mass deaths at Jack's Birthday Site. In the thickness and makeup of the silty mudstones deposited along the shore and the types of freshwater shellfish inhabiting the lake, Dave found evidence that the water level had dropped, an unmistakable sign that the climate had become appreciably more arid, at least for a while. Another possible source of mortality is disease, in particular, any of several ailments that are associated with aquatic environments and that act quickly, such as botulism. We'll never know the whole death story. As I explained in chapter 3, we could easily be led astray were we to forget that the dinosaur bones preserved today probably represent only a small and biased fraction of the dinosaurs that actually lived in the area. One way of explaining our research at Landslide Butte and the Two Medicine River is that we found many bone beds associated with floodplain water holes and small lakes. But another, equally accurate description is that we learned that vertebrate remains stand a greater chance of being buried and preserved near water holes and small lakes, which is an important discovery, because it tells us which sedimentary settings are most likely to contain dinosaurs. As far as Jack's Birthday Site is concerned, all that we can say with complete confidence is that a diverse group of dinosaurs gathered there and that all members of certain subgroups perished at the same time. Despite our uncertainty about the cause of death, the fact that the animals did indeed die in that particular depositional environment is noteworthy, if for no other reason than it made them accessible to us, 74 million years afterward.

So rich was Jack's Birthday Site that we returned to the quarry to dig for more specimens during the summer of 1990. We also investigated a microsite, located near the confluence of Badger

Creek and the Two Medicine River, where Bob had found the remains of frogs and lizards and a clutch of *Troödon* eggs during a trip to Cut Bank the previous winter. Peter Nassar, an undergraduate student at Amherst College in Massachusetts, and a small crew conducted a taphonomic survey of the tiny bones at Bob's Vacation Site. Carrie and I were assisting Peter in the study when we uncovered a beautifully preserved skeleton of a dromaeosaurid dinosaur, the same family of theropods to which *Deinonychus* and the *Velociraptors* of *Jurassic Park* and *The Lost World* belong. Our specimen lacked a skull and was only about three feet long, whereas *Deinonychus* grew to about ten feet, and *Velociraptor* six. We now believe that our specimen is that of *Saurornitholestes*, a small theropod also known from Alberta. Additional discoveries in 1990 included another outcrop that contained abundant skeletal fragments of hadrosaurs, theropods, and pterosaurs. We excavated a small portion of the second multispecies bone bed, enough to permit Dave Varricchio to compare the specimens and sediments there with those at Jack's Birthday Site, which is located on roughly the same geological level. Well into our sixth season on the Blackfeet Indian Reservation, we were still collecting valuable information about dinosaur behavior and evolution.

Late in August, however, our fortunes shifted again, though in this instance as a consequence of entirely new and disturbing forces. We had gone to visit Dave Weishampel and his crew, who were excavating a multispecies bone bed similar to Jack's Birthday Site located in the younger St. Mary River Formation about twenty miles to the northwest. Upon returning to Badger Creek several hours later, we discovered that someone had been digging at Jack's Birthday Site while we were away, someone with a liberal idea of ownership and little appreciation for the correct way to remove fossils from the ground. Equipment was missing and, worst of all, several specimens had been destroyed. Though all of us were stunned by what we saw, Dave Varricchio was particularly upset, which was hardly surprising, given the investment he had made in the quarry.

Excavation at Jack's Birthday Site in 1990. (*Bruce Selyem, reproduced courtesy of the Museum of the Rockies.*)

Having completed almost two years of a three-year-long research project, he would be forced to begin again, somewhere else, if the site were substantially altered. Dave's dissertation depended on the reliability of the information he extracted from Jack's Birthday Site, and that, in turn, depended on the care and thoroughness with which he analyzed the condition of the specimens and measured their location and orientation with respect to each other and to the surrounding sedimentary rock. Once the bones or beds were disturbed they could never be restored to their original state, and any information they might have contained would be lost—forever. This was more than a simple case of vandalism, then. It was an assault on the integrity of our expedition.

Earlier in the season we had received warnings of impending trouble but I never imagined it would go this far. Carrie and Jason were accused of trespassing, even though we had federal permits, a letter of approval from the tribal council, and permission from particular owners and lessees. To make matters worse, someone told the ranchers on whose land we were camping and collecting that we were leaving their gates open and poisoning their cattle, and, most damaging of all, that instead of transferring the fossils to the collection at the Museum of the Rockies at the end of each season, we were actually selling them and pocketing the earnings. Who exactly was behind these attempts to sabotage our work, to say nothing of our hard-won relationship with the tribe, is less important than the profound change in circumstances their actions represented.

In a word, the market in dinosaur fossils had started to boom. A skull that a decade earlier would have sold for one thousand dollars now sold for ten times that amount. A complete skeleton of one of the larger, more popular species might fetch hundreds of thousands of dollars from certain museums. It was only a matter of time before a commercial collector, having heard about our discoveries, would offer to pay the Blackfeet for fossils they or members of the tribe found on the reservation. And when, in 1990, one did, the tribal council accepted, taking advantage of a much-needed source of income at the first opportunity. The decision affected us in two ways, neither of them desirable. On the one hand, further exploration on the reservation was preempted. It was assumed by all parties that the commercial collectors now held first rights to any areas not already staked out as research sites. On the other, and most troubling of all, the areas where we had in fact established a presence were no longer entirely our own. At Jack's Birthday Site, for instance, commercial collectors felt free to dig in any part of the quarry where we happened not to be working at the time, which led, as you might imagine, to more than a little tension on that everbountiful and once-peaceful ridge.

By the end of the 1990 season it had become clear that the era of unhampered exploration on the Blackfeet Indian Reservation was over. During the following winter I met with the commercial collectors to ask them to curtail their operations at Jack's Birthday Site until Dave completed his research. They agreed. Afterward, having fulfilled their promise, the collectors brought in bulldozers and strip-mined the ridge, a sudden, dismaying conclusion to an unusually fruitful dig. Whatever disappointment I experienced at the time, however, was more than offset by the deep gratitude I felt, and still feel, toward the tribe for having allowed us to collect on the reservation as long as we did. For all that the Willow Creek anticline had produced, it didn't begin to compare with the volume and variety of our discoveries at Landslide Butte and the Two Medicine River—the many new species we found, the bone beds and nesting horizons, the sweeping view they afforded of the lives of dinosaurs along the upper reaches of the coastal plain during the late Cretaceous. Most important, from my present vantage, we had found enough evidence to reconstruct not only several persuasive stories of dinosaur social behavior—involving, most notably, mass deaths among herds and parental care in colonial settings—but the outlines of what would become an evolutionary saga.

After realizing that the three new horned dinosaurs from Landslide Butte and the new gryposaur from the Two Medicine River probably fit into evolutionary series, I surveyed all of the dinosaurs that had been found throughout the late Cretaceous period, focusing in particular on the relationship between the appearance or disappearance of certain species and the position of the Western Interior Cretaceous Seaway. That, coupled with Ray Rogers's re-creation of large-scale environmental events along the coastal plains, led me to believe that the evolutionary changes that took place among the dinosaurs living on the upland plains corresponded to changes in sea level. With each new pulse of the sea, the dinosaurs had responded, either by going extinct or evolving at a slightly faster rate. It was as if I were watching an exceedingly slow, sometimes deadly waltz, unfolding on a stage of epic proportions. I could now see the overall pattern the dinosaurs and their environment had followed, the pitch and cadence of their relationship. But I couldn't quite hear the music. That would be the next step—to try to identify the mechanisms that bound the dinosaurs to the rhythms of the sea.

7

ANOTHER LOOK
AT THE
LOWLANDS

"I just can't figure you boys."

The comment came from a good-natured old rancher whose rambling spread encompassed a fossil-rich section of the Milk River badlands northwest of Havre. Of all the foolishness he had witnessed over the years, he evidently had never encountered the peculiar brand that Bob Makela and I were practicing on his land, with his permission if not his full comprehension, one hot summer afternoon in 1972.

"I've got a pile of bones over there you're welcome to," he offered, nodding in the direction of a not-too-distant hill. "Be a lot easier tossing them in the back of your truck than spending all your time digging in this damn rock." I set my pick to the side and removed my hat. Using an already overused shirtsleeve, I wiped away the sweat that had gathered on my neck and forehead. Well, I thought, he's at least half-right; the work would be easier. But it also would be beside the point, since the pile he had in mind contained bison bones, whereas the creatures Bob and I pursued were somewhat older and wilder, which I explained, again, though I'd come to suspect there was nothing I could say to this practical-minded rancher that would make sense of what we were doing. In his eyes, one bone was as worthwhile as another, which is to say, about as worthless as a thing can be and still be something.

That was twenty-five years ago, a much simpler time, at least as far as digging dinosaurs was concerned. (It is a rare rancher today who does not appreciate that some bones are more than what they seem, and some bone hunters better equipped than others to compensate them for that crucial distinction.) I had been attending the University of Montana in Missoula, studying geology, biology, anything connected to paleontology. In 1972, after taking every relevant course offered on campus, I pronounced my formal education completed—a judgment in which, I confess, I was quite alone at the time. Over the strenuous objection of my dean, who held the conventional and admittedlly reasonable view that one should not leave school before meeting all of the requirements necessary to secure a degree, I returned to my hometown, Shelby, and joined my brother in running my father's sand-and-gravel business. About the same time Bob, whom I'd met at the university, took a teaching post in Rudyard, located midway between Shelby and Havre, along the Montana high line, the series of small towns strung across the extreme northern part of the state from the Rocky Mountain Front eastward. Whenever possible, and especially during the summer months, Bob and I explored the Milk River badlands between Rudyard and the Canadian border. The area lay at our back door, after all, and the local landowners readily granted access.

Though back then we certainly could distinguish Pleistocene bison from Cretaceous dinosaurs, in many other ways we were as dumb as the rock we quarried, improvising those aspects of our endeavor in which we had little or no experience, and such aspects were plentiful. Consider, for example, the ungainly contraptions we first used to transport our gear and supplies. We built shallow wooden boxes—miniature coffins, in effect—to which we affixed canvas shoulder straps, allowing us to carry the boxes on our backs. The qualities that one might expect of a modern alloy frame pack—lightness, comfort, flexibility—were entirely absent in our handmade versions. Our gear and supplies included standard paleontological items like burlap, water, and glue, along with such nonstandard items as Vaseline, which, for a time, I employed as a separator, though without much success. One box—Bob's, usually—was reserved for beer, two cases of it, in the understanding that he who was willing to bear so heavy a load during the first part of the day would find his burden steadily reduced, in truth, all but eliminated, as the sun climbed ever higher, sucking the moisture out of everything in sight—air, grass, soil, bodies, especially bodies. Bob and I never went thirsty for long. As I say, a simpler time.

And a time of undiluted pleasure. I found my first dinosaur bone when I was eight years old, on my father's former ranch outside Shelby, and though Dad had sold the place before I was born, fossil hunting kept me preoccupied throughout childhood. Later, while attending college, I collected and studied ancient fish. During that time I also discovered a *Plesiosaurus,* the large, long-necked marine reptile that shared the lukewarm waters of the Western Interior Cretaceous Seaway with the mosasaurs. But it was while exploring the Milk River badlands with Bob in the early 1970s that I decided I wanted to turn my paleontological interests into something more than a pastime. We were still amateurs then, to be sure, but amateurs on a mission, and we were finding lots of dinosaur specimens—teeth, for example, and bone fragments. Then, one day, we stumbled across a partial skeleton of an adult duck-billed dinosaur: our first, and an unforgettable personal milestone. What we didn't discover, however, were eggs, embryos, or babies. In fact, it wasn't until after I left Montana in 1975 to take a preparator job at Princeton University that I collected my first egg. On vacation the following summer, I asked my father to show me once again where his old ranch was located, and while there I discovered what turned out to be the first whole dinosaur egg found in North America. But during the years Bob and I had collected in the Milk River region, scrutinizing mile after mile of outcrops, we had seen nothing of the kind, not even a hint.

Imagine my surprise, then, when Vicki Clouse showed me a handful of eggshell fragments she had gathered along the Milk River, in the vicinity of Fresno Reservoir. It was 1991. Vicki had moved to Havre to take a job at Northern Montana College, in her spare time exploring the badlands northwest of town. Before long she also collected baby bones and squashed eggs, in other words, evidence of a dinosaur nesting ground—in an area I thought I knew like the back of my hand.

Why had Bob and I not detected similar evidence? Certainly our search image played a part. We had never seen egg fragments before and, what's more, we weren't expecting to—a serious handicap, since they are very difficult to detect even under the best of circumstances. The vagaries of erosion, too, may have contributed. When Bob and I searched the Milk River badlands, the layers containing egg material might not have been sufficiently exposed yet. (This is the feature, of course, that makes paleontology a field of continuous discovery: as sedimentary rock weathers throughout the world— and it is always weathering—new fossils are inevitably exposed.) Finally, the oversight might have been a matter of dumb luck, a right turn instead of a left, scouring one side of a hill instead of the other, ending a search prematurely.

There was a second, more important reason for my surprise at Vicki's finds. The Milk River badlands belong to the Judith River Formation, which, as you may recall from the geological description of north-central Montana in chapter 4, is separated from the Two Medicine Formation by the Sweetgrass Arch and represents terrestrial sediments deposited along the lowland plains as the seaway retreated, then, 75.4 million years ago, reversed course and expanded for the last time. And as I explained when discussing fossil preservation within different depositional environments in chapter 3, the gray, brown, and sometimes black sediments of lowland areas, which represent poorly drained swamplike or seaside settings, tend to be highly acidic. More often than not, dinosaur eggs do not survive in such a geochemical context; rather than being fossilized, the calcium shell dissolves.

I should add that my friend Ken Carpenter, of the Denver Museum of Natural History, first proposed this scenario; by contrast, based on what I saw in the Two Medicine Formation, I had conjectured that dinosaurs laid their eggs only in upland regions, far away from the seaway. One might find bones in the lowlands, I hypothesized, but nothing that would indicate nesting behavior. So whereas in the early 1970s I didn't look for dinosaur eggs in the Judith River Formation because I had no reason to do so, twenty years later I had no intention of looking because I had every reason to believe there was nothing to find. Vicki proved me wrong. The lowland sediments are more complex geologically than I had surmised. Better yet, her discovery of eggs and babies cut a trail into new territory, which in truth was old territory for me, but viewed through different eyes. Remember that by 1991, our work within the boundaries of the Blackfeet Indian Reservation had been curtailed, save for one modest excavation, Bob's Vacation Site, where the specimens were too small and damaged to be of interest to commercial collectors. The eviction helped me to make a decision—to expand my search to other strata from the Cretaceous period, comparing any fossils they contained with those we had found at the Willow Creek anticline, Landslide Butte, and the Two Medicine River. Now I had a promising starting point—the lowland Judith River Formation.

Our first excavation in the Milk River badlands actually took place at a site discovered by another amateur, John Bruninga, of Kalispell. While hiking through the area in 1992 John found a clutch of whole, unusually white eggs, about the size of silver dollars and, we eventually discovered, each containing a tiny embryo. Nearby he also found a few bones of some variety of duck-billed dinosaur. John brought the eggs to Vicki, who had by then entered the graduate program at Montana State University in Bozeman, working on her master's thesis. She couldn't identify the creature that laid the eggs. Nor could I, except to say that it wasn't a dinosaur. Only recently did we learn that the clutch belonged to a Cretaceous freshwater turtle. In 1992, however, the most important feature of John's eggs was not the identity of the animal that laid them but the fact that they had been preserved in those particular lowland sediments for some 75 million years. Perhaps dinosaur eggs had survived as well. Under Carrie Ancell's supervision the crew did indeed find eggs—a partial clutch belonging to an unknown species of lambeosaur. They also unearthed lambeosaur and hadrosaur skeletal fragments, but, unfortunately, no telltale skulls.

Although the exact identity of the dinosaurs at Egg White Site, so-named because of the color of the turtle clutch first found there, was a mystery, this initial foray into the Milk River badlands proved encouraging. We now had two duck-billed-dinosaur nesting grounds, counting the yet-to-be-excavated area where Vicki had uncovered the first eggshell material. To be sure, I didn't expect the Judith River Formation to yield the volume and array of fossils that came out of the upper layers of the Two Medicine Formation; as wrong as I had been about the likelihood of finding eggs in lowland sediments, it still remained exceedingly unlikely that a bone bed comparable to the maiasaur mass kill at the Willow Creek anticline or a nesting site as extensive as the hypacrosaur rookery near Landslide Butte would have survived in anything remotely resembling their original forms. What the acidic soils didn't destroy, the many large and meandering streams common to lowland environments would have washed into the sea. Nevertheless, the fossils that the sandstones of the Judith River Formation did yield were precisely those I sought—aggregates of eggs, embryos, and babies.

During the fall of 1992 I secured a grant from the National Science Foundation to continue exploring the Milk River badlands, as well as several other parts of Montana, each representing a different section of Cretaceous rock. Beginning in 1993 and lasting three years, the NSF study would focus on life history strategies, a term biologists use for characteristics, both behavioral and morphological (which includes anatomy and physiology), that have evolved for the purpose of survival. In the interest of simplicity, then, let's refer to them as survival strategies, further picturing the phrase as a clearer and more inclusive way of talking about features of dinosaur life that we had been studying, though less systematically, since 1978, when Bob Makela and I found the first *Maiasaura* nest at the Willow Creek anticline. Typical morphological strategies, for example, include egg size, the number of eggs in a clutch, the size of the hatchlings at birth, their subsequent growth rates, and the size at which juveniles left their nests or nesting grounds. The behavioral strategies that interested me most fall under the heading of socialization—gathering in groups to reproduce and raise young, forage, or hunt.

Having appeared and persisted because they offer certain evolutionary advantages, most survival strategies also carry liabilities, depending on circumstances. Consider, for instance, an altricial bird that lays one egg from which emerges a hatchling that cannot fend for itself but grows rapidly. The advantage of this strategy is that the adults can devote all of their energies to feeding and protecting their sole offspring, increasing its likelihood of survival. The risk is that if a predator manages to outwit the adults and kill the hatchling or if it succumbs to illness, the parents will have lost their only chance to reproduce. A precocial bird that produces a dozen hatchlings, on the other hand, cannot give them the same degree of attention, which places them in greater peril than the single altricial baby. But precocial hatchlings, even if they grow at a relatively slow rate, possess the advantage of being able to take care of themselves from birth or very shortly afterward. And in this instance whatever they lack in size they make up for in numbers; several, even most, can die and the parents' genes will still survive in the lucky juveniles that live to sexual maturity and reproduce.

Practically speaking, our effort to survey the survival strategies of dinosaurs inhabiting the coastal plain throughout the Cretaceous period meant that exploration and excavation would be occurring at several locations simultaneously. Among the areas we expanded to after starting digs in the Milk River badlands was the early Cretaceous Cloverly Formation in south-central Montana. Comprised of terrestrial sediments—claystone and sandstone, mostly—deposited along the uplands 120 million years ago, the Cloverly was one of the formations in which the skeletons of two kinds of dinosaur had been found together in a manner that suggested a strong association between them.

The first species is *Deinonychus,* the carnivorous theropod that John Ostrom used to make the argument that dinosaurs gave rise to birds. At Rainbow Butte, the name sometimes used for the site where in 1964 Ostrom initially excavated the *Deinonychus* bones, he eventually uncovered the remains of as many as five different skeletons. More intriguing, resting within the group were the tail and several other parts of a single *Tenontosaurus,* the plant-eating ornithopod that we had found further north in the Cloverly Formation, at Devil's Pocket. Ostrom speculated that the tenontosaur was attacked by a large number of the theropods and managed to kill five of them before perishing in the struggle. In other words, *Deinonychus* hunted in packs—one of the survival strategies I had targeted. But Ostrom hadn't removed all of the bones from the site; the overburden had become too imposing. When I called him at Yale University to ask if he had any objections to my sending a crew to Rainbow Butte, he replied that he'd be delighted to see his original quarry completed.

Meanwhile, Desmond Maxwell, a postdoctoral colleague from the New York College of Osteopathic Medicine, who was working with me at the Museum of the Rockies, had been exploring in the vicinity of Rainbow Butte when he discovered eggshell fragments and baby bones. In 1993 Des led a crew back to Rainbow Butte to excavate the eggshell site and to reopen Ostrom's historic quarry. The bones, we learned, belonged to a small ornithopod, *Zephyrosaurus,* a close cousin of *Orodromeus.* The deposit itself turned out to be a microsite, the eggshell and skeletal remains having been washed in from elsewhere. No evidence of a nesting horizon was uncovered. As for Ostrom's quarry, at the very least, the *Deinonychus* bones represented some kind of social group. I wanted to identify other characteristics of its members, including their age. Was social behavior restricted to adult dinosaurs or did juveniles participate as well? We also were looking for more tenontosaurs. A few years earlier Catherine Forster, then with the University of Pennsylvania, had examined two sets of specimens at the American Museum of Natural History that had been collected in Montana in the 1930s and 1940s, concluding that each assembly included several juvenile tenontosaurs. Forster suggested that in the interest of survival the young ornithopods formed social groups, perhaps shortly after hatching. The Cloverly Formation seemed like a good place to test that notion.

Intermittently over the course of three seasons, we examined the chalk-white and brick-red layers of the hundred-foot-high, pyramid-shaped promontory Rainbow Butte, as well as the surrounding badlands, unearthing in separate locations two new tenontosaurs, a baby and an adult, but no groups. The adult was especially interesting. For one thing, it was almost complete, from nose to tail; only its hands and one of its hind legs were absent. For another, where the missing parts should have been we found *Deinonychus* teeth, many more of them than could have been lost by a single animal. What's more, the tenontosaur's rib cage was unusually distended, suggesting that it had been ripped open, that an animal had been trying to get at the viscera within. It appeared that at least four of the theropods had been feeding on the tenontosaur, which they may have killed, too—a second instance of pack behavior in the same area. The discovery added to my doubts about Ostrom's interpretation of the first *Deinonychus-Tenontosaurus* scenario. Though standing no more than four feet tall and weighing only 130 pounds, *Deinonychus* was a fearsome predator (imagine an animal about the size of the *Velociraptors* in *Jurassic Park*). For a lone plant-eating tenontosaur to kill five of them at once would be the equivalent of a wildebeest slaying a similar number of attacking lions.

The fossil evidence, too, raised questions about Ostrom's original interpretation. Most of the *Tenontosaurus* skeleton was nowhere to be found, whereas the *Deinonychus* skeletons were more complete and a couple of them had articulated feet. If geological processes—erosion, in particular—had played a part in disturbing the remains after the flesh had disappeared, one would expect a very different pattern: the *Deinonychus* bones, being much smaller and lighter than those of *Tenontosaurus,* would have been washed away first. If, on the other hand, the pack had been feeding on the tenontosaur carcass, parts likely would have been removed, even consumed elsewhere, just as sometimes a pride of lions will, in the course of eating its prey, separate limbs, then carry or drag them some distance from the kill. It's safe to assume, I think, that a *Deinonychus* pack was feeding on the tenontosaur when five of its members died for reasons that we can't yet determine. Whether Ostrom's group, or, for that matter, the one we found, actually killed its tenontosaur, however, remains unclear; the taphonomic evidence is ambiguous. That question aside, the theropods had not yet reached adulthood when their lives ended. Microscopic studies of the internal structure of the *Deinonychus* bones revealed that the animals were still growing.

From the hunting packs of the Cloverly Formation to the herds and colonies of the Two Medicine Formation, social grouping appears to have been a common survival strategy among many kinds of dinosaurs inhabiting the coastal plain of North America during the Cretaceous period. But was this true as well of earlier times, among species that predated the small theropods of Rainbow Butte, the hadrosaurs and hypsilophodontids of the Willow Creek anticline, the lambeosaurs and ceratopsians of Landslide Butte? We hadn't set out to address this question; events in the field led us to it. We repeatedly came across older rock, the Jurassic sediments of the Morrison Formation, that contained dinosaur remains, despite the common belief that dinosaurs could not be found in such sediments in Montana. Actually, my detour into the Jurassic period had begun as far back as 1990, when we excavated a partially intact skeleton of

The Livingston Sauropod Site (*from left to right*): Jason Horner, Donna Coles, Allison Gentry, and Hilory Korte (*far right*). (*Bruce Selyem, reproduced courtesy of the Museum of the Rockies.*)

a two-thirds-grown sauropod near Livingston, in the south-central part of the state. We later learned that the dinosaur probably represents a new species, one that may well be a precursor of *Diplodocus* and *Apatosaurus*, both of which it resembles in critical ways.

Diplodocus and *Apatosaurus*. The mere thought of them—more to the point, of removing them from the ground—makes me shudder. Fully grown, *Diplodocus* was about ninety feet long and weighed twelve tons; *Apatosaurus*, seventy feet and twenty-five tons. Excavating the skeletons of these immense creatures is an extremely laborious process, and that for the most part is what it means to hunt for dinosaurs in Jurassic sediments—uncovering and removing sauropods, by far the most common specimens from the period. As I explained earlier, single skeletons offer little if any insight into the issues that concern me most—group behavior and evolutionary processes that can be understood only by comparing large numbers of fossils—and when the dinosaur is very large the amount of effort required to unearth a single skeleton so greatly outweighs its scientific value that I try to avoid such excavations altogether.

I don't always have a choice, however. Toward the end of the 1991 season the Bureau of Land Management asked us to salvage an allosaur at a site near Howe Quarry in Wyoming.* While Bob Harmon and his crew were removing the last section of that large Jurassic theropod, they discovered a sauropod buried beneath it. The animal, evidently a species of *Diplodocus,* was not fully grown when it died, a desirable feature, from my perspective at least, and its skeleton had been well preserved, another, more obvious advantage. Since the Jurassic period is underrepresented in the fossil collection at the Museum of the Rockies, I felt obliged to remove the skeleton and bring it back to Bozeman for preparation and public display. During the summers of 1992 and 1993, Bob, Carrie Ancell, and other crew members continued digging at the site. One day toward the end of the excavation they found a rib and vertebra of another sauropod, this one lying on its back. The rib alone was nine feet long. Soon they uncovered a second pelvis as well, and it measured five feet by five feet by five feet. Trying to picture the skeleton to which this enormous bone belonged, I very quickly realized that it was much too large to house at the museum. "Take out the rib and the vertebra," I told Bob, "then cover it up. I don't want to see the rest of this animal." Judging from the expression of relief that flashed across Bob's face, I'd have to say that he'd seen enough, too.

Not every sauropod we encountered during our unplanned journey through the Jurassic period was a solitary colossus, however. In 1992 Peter Fonda introduced me to his sister, Jane, and her husband, Ted Turner, and I took the couple on a paleontological tour of one of the Turner ranches outside Bozeman. My assignment was to char-

---

*For more information about the Howe Quarry excavation see the afterword, which deals with the controversial issues raised by commercial collection on public lands.

acterize outcrops and to construct a geological narrative for the region while speculating on the fossils that might be found there, an exercise that proved highly enjoyable because within the ranch there is a wide variety of exposed strata, including a promising section of the Morrison Formation. Before long I discovered the neural spine of a dinosaur vertebra, then a handful of toe bones, all of which turned out to belong to an *Apatosaurus* bone bed. Best of all, it was a juvenile bone bed, the first of its kind anywhere. Supported by a grant from Turner, I immediately assembled a crew and dispatched it to T&J Site, as we called the quarry, to dig up as many of the young apatosaurs as possible. Unfortunately, the surrounding rock matrix is an extremely hard limestone. Rather than using the standard tools—picks and shovels to remove overburden, whisk brooms and chisels for the more painstaking detail work—the crew relied almost exclusively on diamond saws, carving out whole blocks of limestone containing skeletal remains. Back at the museum laboratory, each block then had to be acid prepared, that is, bathed in a solution caustic enough to dissolve the matrix but too weak to damage the bones within, an exceedingly slow process. In 1994 I suspended operations at the quarry until the specimens we had extracted by then could be prepared, a task that continues to this day.

Although T&J Site closed down before we could identify the probable cause of death or do much in the way of determining precisely the original environment in which the deaths and subsequent deposition took place, the fact that a group of juvenile apatosaurs died in the same place presented me with the first opportunity to study among sauropods a survival strategy that previously I'd observed only in more recent species of dinosaur. Paleontologists have found trails of fossil footprints, called trackways, elsewhere in the West, suggesting that at least some of these plant-eating behemoths may have traveled in small herds, most likely for protection from roving predators, but in such aggregates juveniles, subadults, and adults appear to have been present. The fossil assemblage at T&J Site, by contrast, raised the possibility that certain sauropods formed groups based on age, though if the purpose for doing so was

Sauropod bone bed excavation on the Turner Ranch (T&J Site).
(*Bruce Selyem, reproduced courtesy of the Museum of the Rockies.*)

not protection it is hard to imagine an alternative explanation. That was about all I could say on the subject of Jurassic juveniles until the fall of 1994, when Curt Padilla came to see me.

Curt is a dedicated amateur collector who has been a friend of the museum for many years, stopping by whenever he and his family find dinosaur specimens he thinks might be of interest to us. This time Curt and his wife had been exploring near Rainbow Butte, where Ostrom's *Deinonychus-Tenontosaurus* quarry is located, when they came across a couple of sauropod bones. Since they'd made their discovery on the second Sunday in May, the Padillas christened the spot Mother's Day Site. I asked Bob Harmon to return with Curt to the area and conduct an initial investigation to determine whether a full-scale excavation was warranted. Bob reported that the sediments were part of the Morrison Formation and that relatively little overburden stood in the way of collecting additional bones. Then he offered the most intriguing observation of all—the skeletal remains that were evident near the surface seemed to be monospecific and segregated by age. Bob's assessment proved correct. The following season, when Kristi Curry took a crew to the site and started digging, they found nothing but young sauropod bones, hundreds of them, all from individuals about the same size (which, interestingly enough, is roughly equivalent to the skeletons at T&J Site), and all apparently belonging to a single species—*Apatosaurus*.

Mother's Day Site seemed too good to be true: a second segregated group of Jurassic juveniles that, unlike the first, was entombed in Morrison mudstone, a much more tractable matrix than Morrison limestone. The mudstone, in fact, provided a clue to the type of environment in which the apatosaurs died and, possibly, to the manner of their death. Mother's Day Site, which is about a

Mother's Day Sauropod Site in southern Montana. Celeste Horner (kneeling at computer) and Kristi Curry map skeletal elements using a sonic digitizer. (*Bruce Selyem, reproduced courtesy of the Museum of the Rockies.*)

quarter mile long, borders a unit of marine sediment called the Swift Sandstone. It appears, then, that the sauropod bone bed is located on what was once the muddy shore of an ocean—specifically, the seaway before the three expansions of its western boundary that took place during the Cretaceous period. How did so many bones of the same dinosaur at the same stage of development end up in that particular environment? The answer may rest in part with the kinds of skeletal remains Kristi and her co-workers uncovered at the site. Typically, sauropod skeletons are found in stream deposits, where, also typically, small bones (the toes, for instance) are very rare, presumably because they were washed away by the stream while the fine-grain material carried by the stream slowly buried larger remains such as leg bones, neck vertebrae, and skulls. Mother's Day Site, however, is not an ancient streambed but a mud bank. It's also unique in that it's chock-full of small bones, but mostly from the lower part of the apatosaurs' bodies. Feet and legs are plentiful; shoulders, ribs, and hips uncommon; neck vertebrae rare; and skulls so far absent altogether.

Before considering the implications of this unusual pattern, it might be helpful to review social groupings among Cretaceous dinosaurs, especially what we have learned from the many monospecific bone beds of duck-billed and horned dinosaurs we discovered in the upper layers of the Two Medicine Formation. For one thing, the mass deaths were caused by either volcanic fallout, flood, or drought (including such drought-induced factors as disease and opportunistic predation). At Mother's Day Site no evidence exists for any of these scenarios. For another, the gigantic herds of Cretaceous plant eaters appeared on the coastal plain precisely when deciduous plants were well on their way to colonizing all of North America, along with the rest of the continents. Since they replaced their leaves each year, the plants would have provided the herds with an abundant and endlessly renewable source of food, as long as the herds of duck-billed and horned dinosaurs migrated with the seasons. But there were no deciduous plants during the late Jurassic, when the Morrison Formation was deposited, no readily replenishable food source that could have sustained a large group of extremely large plant-eating sauropods.

I'm still a bit puzzled by Mother's Day Site, but in keeping with the principle of parsimony—the idea that the simplest, most economical explanation is the best explanation—I tend to think that if the remains of juvenile sauropods are deposited in the same location but there is no discernible reason for the sauropods to have been together when they were alive, then we can't say they represent a social group, though that remains a possibility. Aggregates can form in other ways, after all, especially those that appear in the fossil record. Take the sauropod trackways I mentioned earlier. Some paleontologists are reluctant to accept them as evidence of herding because there is no way to tell whether they were made by a group traveling together or many individuals traveling separately. A similar accumulation, eliciting an equally ambivalent interpretation, might have occurred at Mother's Day Site.

Here's a likely scenario: The sauropods were trapped in the near-shore mud. Perhaps the deaths occurred one by one, over a period of time, with the bones that were buried—the lower parts of their bodies—being preserved and the rest being washed away. Imagine an elephant wandering into a patch of thick mud. Flat-footed and unable to lift its heavy legs very far off the ground, it could easily get stuck, permanently stuck. Sauropods, being larger and more cumbersome, would have had a more difficult time extricating themselves. As for the uniformity of age among the apatosaurs at Mother's Day Site, it's possible that, depending on the depth, extent, and viscosity of the mud, only sauropods of a certain size would not have been able to walk out of the trap on their own. Even if the hapless animals later were besieged by predators, their feet and the lower bones of their legs would have remained in the mud, out of reach, and thus fossilized in place. Someone could argue, of course, that a group of sauropods that perished together would produce the same pattern, which is true. In the absence of additional information we cannot say whether they died separately or as a group. For now, though, the evidence, such as it is, points to trapping, which is a form of mass death that has been documented in mammals and some other animals but which no one had seen among dinosaurs before.

What did our side trip into the Jurassic period accomplish? From my standpoint the most significant finds were T&J Site and Mother's Day Site. Taken together, the two bone beds raise the possibility, and, I must emphasize, only the possibility, that sauropods were gregarious and, moreover that they may have segregated themselves on the basis of age. If this eventually proves to be the case, then social grouping as a survival strategy emerged early in the evolution of dinosaurs, at least among plant eaters. That said, however, the best places to study social grouping, at least in Montana, remained the same: sedimentary outcrops from the Cretaceous period, because the strategy seems to have become widespread by then, in terms of both the range of dinosaurs that employed it and its geographical distribution on the coastal plain. Further exploration of the Milk River badlands outside Havre, for instance, continued to turn up evidence that duck-billed dinosaurs nested in colonies in the lowland Judith River Formation, though the exploration there was not without disappointment.

During the 1992 season, before the excavation of Egg White Site was completed, Carrie and Bob, who had been supervising the removal of the surprise sauropod near Howe Quarry, explored the area where Vicki had first collected eggshell fragments. Early in their scouting trip the group experienced one of the inevitable yet unpredictable episodes that remind paleontologists that fieldwork consists of more than the search for fossils. A rattlesnake bit Carrie on her ankle. Bob ran to get his truck, parked about a mile away. But by the time he got back, Carrie's leg had turned black and blue and had swollen its entire length, from her heel to her hip. More ominously, she was going into shock. They raced into Havre, where Carrie received medical treatment and, thankfully, soon recovered. But she will never forget the ordeal. Nor will we, since it was immortalized, inadvertently, in the name of the place where it occurred: Long Time Waiting. I say inadvertently because Vicki had originally assigned the name in frustration at the small number of fossils found there. Later the crew's luck turned. They came upon a badly weathered partial clutch of lambeosaur eggs, one of which contained an embryo—partial compensation, perhaps, for Carrie's troubles.

We encountered another, more familiar obstacle the following summer—nasty weather, which, once it settled in the Milk River badlands, refused to leave until still worse weather, winter's, arrived to take its place. The season started on a promising enough note, however. Carrie led a full crew in reopening Egg White Site. Though the quarry was small, no more than 120 feet square, it soon yielded hadrosaur eggshell fragments, a partial clutch of eggs, and baby bones. Then the rains came, attended by relentless wind and bone-numbing air, all crowded beneath a heavy gray canopy that stretched from one end of the prairie to the other. Thereafter our work was interrupted so often, not only along the Milk River but at Howe Quarry, Rainbow Butte, Bob's Vacation Site, Egg Mountain, and the sauropod site outside Livingston, as well as a triceratops dig in eastern Montana I haven't mentioned before, that the season seemed to consist of nothing but interruptions. To oversee these widely dispersed operations I had planned to keep moving, driving from one site to another, but instead spent most of the summer either rained in or rained out, everywhere but where I needed to be. The final insult took the form of a violent storm that blew away the entire camp—tents, gear, everything—at Egg White Site. Carrie and her demoralized crew were forced to shut down the excavation, the first time that has happened in all my years of paleontological fieldwork.

The 1993 season was not entirely discouraging, however. Volunteers Barbara Lee and her son Robbie, who have contributed generously to our research, found a hillside littered with eggshell fragments and baby bones only a couple of hundred yards from Long Time Waiting. Upon returning in 1994 to excavate the find, which Robbie named Eggroll Site, the crew, consisting of Carrie, Vicki, and a group of six or so volunteers, made the single most exciting discovery of our several seasons in the Milk River badlands. The excavation was difficult, lasting the entire summer and turning up specimens at an agonizingly slow pace—about one hadrosaur bone or bone fragment per day. But the skeletal remains were small, belonging to juveniles of various sizes and very likely indicative of a nesting horizon, so I instructed the crew to continue digging.

Then, one sweltering afternoon in August, they uncovered a large hadrosaur egg, about the size of a football, which they promptly removed, only to discover that several others lay alongside it. Normally we find eggs weathering out of sediments, making it impossible for us to determine whether the clutches we retrieve are complete. This excavation was different. The crew had quarried several feet of rock from the top of the hill before reaching the clutch, which means that it had never before been exposed, never disturbed by erosion. In other words, we could be certain that it was intact. Further digging revealed a nest about six feet long and four feet wide, within which twenty-two eggs were arranged in an oval pattern. Having slogged through the endless rains of 1993 and fried in the merciless heat of 1994, the crew was thrilled at this turn of events—and justifiably so. They had found the world's first complete clutch of duck-billed dinosaur eggs.

Eventually Vicki and the others uncovered several other clutches in the Milk River badlands. Along with the specimens found at Egg White Site, Long Time Waiting, and Eggroll Site, all were located on three distinct nesting horizons, two of which are quite extensive—up to a mile in length. There's no question in my mind that lambeosaurs repeatedly returned to that section of the Judith River Formation to establish nesting colonies. Which part of the lowland plains the section represents is less clear. If we knew the age of the section then we'd also know where it was in relationship to the retreating or expanding edge of the seaway. But to date the strata precisely, we need a layer of volcanic ash, which contains argon, an element that naturally decays at a fixed rate into generation after generation of radioactive isotopes. By measuring the

This clutch of eggs found in the Milk River Badlands was laid by a crested duck-billed dinosaur. (*Bruce Selyem, reproduced courtesy of the Museum of the Rockies.*)

ratios of the isotopes to the parent element we can determine how long the argon has been in the ground and thus the age of the surrounding rock. But there's very little volcanic ash in the Judith River Formation.

In the absence of measurable argon, we must rely on less precise methods: comparing the sediments with other strata, in both the Judith River Formation and nearby formations, such as the Two Medicine, which have been dated, and studying the sediments themselves for clues to the kind of environment they represent. At this juncture in her comparative analysis, which she's conducting for her thesis project, Vicki tends to believe that the sediments, and thus the nesting horizons embedded within them, lay near the top of the formation, which is to say that they were deposited fairly late during the final expansion of the seaway. In fact, she estimates that the sea was no more than twenty miles away when the duckbills laid their eggs. But I'm not yet convinced this is the case.

One thing, however, is certain: Though the Milk River sandstone northeast of Havre unmistakably represents a lowland environment, the strata that contain the nesting horizons, which are comprised of silty mudstone, resemble those of upland environments. They are bright gray, for instance, and in some places almost green, which means that they were well drained. This suggests to me that the nesting horizons were preserved because the surrounding sediments were somehow spared by the numerous, widely meandering rivers that we know scoured the eastern part of the coastal plain, erasing most traces of dinosaur existence. Given the swamplike condition of much of the region, it stands to reason that the Milk River duckbills would have favored any location that remained high and dry, so to speak, returning often to lay their eggs. And that's exactly the scenario that the evidence points to at the two major nesting horizons we discovered there.

The exploration of the Judith River Formation is still a work in progress. Since we haven't yet found a skull associated with the eggs, babies, and juveniles of the Milk River badlands, we don't know exactly which duckbills constructed the colonies there.

Even so, nothing we found during the study of dinosaur survival strategies in the Judith River Formation, or, for that matter, elsewhere in Montana during the early 1990s, nullified our original ideas about parent care and other forms of social behavior. Yet a striking new pattern had developed, one that I couldn't have predicted on the basis of our findings at the Willow Creek anticline alone. To be specific, after almost twenty years of exploration, the *Maiasaura* nest Bob Makela and I unearthed in 1978 remains the only one that contained post-hatchling babies. In every other nesting horizon we discovered, from the colonies at Landslide Butte and Blacktail Creek North to the Milk River egg sites, we found plenty of baby and juvenile bones, in some instances thousands of them, but no skeletons inside a nest. Significantly, though, the immature skeletal remains were always confined to the colonies. By the time we left the Blackfeet Indian Reservation, this pattern—baby bones within horizons but outside nests—had become so firmly established that I began to think of it as the rule rather than the exception. Indeed, a little experiment I conducted in 1991 convinced me that I shouldn't expect anything else. It turns out that although our ideas about parental care have withstood the test of further research, the starting point for those ideas, the first *Maiasaura* nest, is an anomaly.

I'll resolve this apparent paradox, as well as describe the crucial experiment, in the next chapter. But first I wish to point out that the reinterpretation of one of the key specimens of the Willow Creek anticline exemplifies how the meaning of paleontological evidence shifts as additional evidence is gathered. As I said earlier, paleontology is a field for those who place a higher value on knowing what is real—practically speaking, a continuous process of revision—than on being right. As historical scientists we cannot conduct experiments on our subjects; they're dead and, to make matters more difficult, the environments they inhabited have vanished. Nor can we derive conclusions from first principles or perform calculations based on universal equations. Instead we examine the evidence, the fossils and the geological contexts within which they are found, then reconsider it, and reconsider it once more, going back again and again, revisiting previous stages in the history of paleontology while at the same time returning to earlier periods in the history of life.

I've already described this method in terms of retracing footsteps, those of other collectors, amateur and professional alike, as well as one's own. But one might also characterize it as an attempt to study something from all angles, a circling approach, very much like that of a hawk as it hovers above a field in search of prey or the downward spiral of a flock of pelicans landing on water. That's how I see our most recent expeditions in the Milk River badlands. I returned to the terrain where Bob Makela and I taught ourselves how to track down and dig up dinosaur fossils—returned, essentially, to my origins as a paleontologist—finding things the second time around that we never dreamed existed there. And given the cyclic nature of my work I suspect that someday I'll go back for a third look, with new eyes, and, if I'm lucky, leave with yet another version of what happened there.

8

FROM EGGS TO EVOLUTION

Immediately northeast of Malta lies Bowdoin National Wildlife Refuge, a thirty-five-square-mile patch of hillocks and marshland that encloses an S-shaped reservoir known as Lake Bowdoin. The lake in turn contains three islands, and every spring American white pelicans, double crested cormorants, and seagulls return to those water-bound sanctuaries to bear and rear their young. One cloudless summer day well into the 1991 nesting season, I was walking along the south shore of the lake when I spotted a section that seemed shallow enough to permit me to wade across to the largest of the islands, an acre or so of barren hardpan bordered by fireweed, scraggly willow, and an assortment of other small bushes. I pulled on a full-length wading suit, the neoprene version of bib overalls commonly used by determined stream fishermen, and set out on an odyssey through the water, toward the flurry of birds presently making the island their home.

Lake Bowdoin is alkaline—that is, salt-laden—and thus mortally hostile to fish and other aquatic creatures, to say nothing of human beings in whom the desire to witness its avian inhabitants firsthand occasionally outweighs the dictates of caution. I am that kind of person and this was just such an occasion. About halfway to the island the mud on the bottom grew so thick and binding that I began to worry: were I to proceed any farther I might find myself trapped, something like the juvenile sauropods at Mother's Day Site, a predicament that would force me to shed the waders and swim back to shore. As it was, simply extracting a foot and taking a step in any direction required most of the strength I could muster. But that wasn't the worst part. The birds on the island weren't altogether pleased to see an intruder coming their way, and given the noise I made during the crossing and my slow-motion pace, every one of them now knew that I was approaching the nesting ground. Whether by virtue of temperament or their superior size I can't say, but the pelicans seemed particularly eager to express their displeasure. They also seemed to understand that there wasn't much a single muck-bound man could do to defend himself against an attack from above.

No sooner did I twist about and start back for shore than I was under siege, beset by a dozen adult pelicans, the very same birds that Bob Makela and I had always considered bearers of good fortune. I can't blame them for coming to the defense of their young, of course. Nor can I help wondering if Bob was not somewhere in the vicinity as well, chuckling at his partner's fateful lack of prudence. These are afterthoughts, however. At the time my attention was directed elsewhere. One by one, sometimes in twos and threes, and from all directions it seems, the pelicans dove at me, their legs jutting downward like descending spears. When they were within a few feet of striking my head, they abruptly stalled, spread their wings to full length (some nine feet on the largest), and opened wide their giant mouths, which unmistakably stank of sour fish. That day I may well have set a new world record for sprinting through mud and murky water. After stumbling several times I finally reached land, where, to my amazement, the pelicans continued their onslaught for another ten minutes while I remained on my knees, crouched over, covering my face with my arms. Nesting on an island in a shallow, soft-bottomed reservoir, I realized then, is a pretty effective survival strategy. The pelicans needn't worry about terrestrial predators, at any rate.

What had prompted me to wade into the middle of Lake Bowdoin? Since leaving the Willow Creek anticline, we'd found two very promising *Hypacrosaurus* nesting grounds—the largest known dinosaur rookery in the Western Hemisphere at Landslide Butte and the fossil-rich Blacktail Creek North colony in the Two Medicine River region—but at neither location, nor anywhere else, for that matter, did we find evidence to suggest that the lambeosaurs remained in their nests after hatching. Although the Landslide Butte site included millions of eggshell fragments and hundreds of baby bones, the bones represented only embryos or very young juveniles. The reason for that, we had eventually learned, was that the animals had suffocated in their eggs, or during hatching, the entire nesting ground having been rapidly buried beneath a thick layer of searing volcanic ash. At Blacktail Creek North, by contrast, we found a great number of baby bones of various size on the same horizon with eggshell fragments but none inside nests per se. Uncertain what these configurations implied about parental care, I decided to compare the dinosaur nesting grounds with those of contemporary birds. In 1991 officials at the Bowdoin National Wildlife Refuge gave me permission to study the cormorants and white pelicans of Lake Bowdoin.

Following my hair-raising first encounter, graduate student and photographer Terry Panasuk and I set up a blind so we could observe and document the activities of the birds without disturbing them. It soon became apparent that the adults were leaving the refuge on a regular basis, flying, we eventually discovered, as far as a hundred miles to other freshwater lakes and rivers to catch fish for their young. On several separate occasions, while the cormorants and pelicans were away, a park ranger transported me by air boat to the two larger islands, where, in relative peace, I conducted taphonomic surveys of the nesting grounds. Taphonomy, you will recall, is the study of animal remains. I mapped the locations of cormorant and pelican bones, their size and orientation, and examined their general condition. I also dug shallow pits and recorded the number and depth of bones located below the surface, as well as how the numbers correlated with the placement of nests. The islands being slightly elevated at their centers, several concentrations of bones had been washed together by streamlets, and isolated skeletal fragments were relatively abundant within the soil as well, all of which was reminiscent of the fossil assemblages at Blacktail Creek North. What's more, carcasses in various stages of decay were scattered at random across the nesting ground.

But the most suggestive finding by far concerned the stages of development represented by the remains. Every specimen belonged to either a baby or a juvenile. I found no evidence of dead adults on the islands, not a single full-grown skeletal fragment or carcass. This is significant because both cormorants and pelicans are altricial; their hatchlings are unable to walk or feed themselves. The pelican chicks are of particular interest because even after they became mobile they stayed in the rookery, confined initially to their nests and later to nursery areas and, at times, sections of the lake immediately offshore of the islands, where they were cared for and protected by a few adults, the rest of the colony off finding food, until they were about three months old and had attained a size almost equivalent to their parents. And as you well know by now, this is precisely the pattern we saw among the new lambeosaur—*Hypacrosaurus stebingeri*—we discovered in the upper layers of the Two Medicine Formation.

Judging from the range of bones we found at Landslide Butte and Blacktail Creek North, the movement of newborn hypacrosaurs was restricted to the nesting ground until the hatchlings had grown appreciably. To my way of thinking, this indicates that the dinosaur was probably altricial. There would have been no reason for the young hypacrosaurs to remain within the colony unless they were being fed, as well as protected, by their parents. In fact, staying put would have been suicidal otherwise, since the plants upon which they depended for survival would have been available only outside the much-trampled, largely barren soil of the nesting grounds. And it stands to reason that if the young were being cared for by the adults, it was because neither of them had a choice—in other words, because the young were incapable of caring for themselves. Those organisms with the most parsimonious survival strategies, remember, usually stand the best chance of perpetuating their kind; more to the point, no parent dinosaur would have placed its own life in jeopardy by squandering its energies on an offspring that didn't need assistance.

After observing the pelicans of Lake Bowdoin in the summer of 1991, I no longer believed that to demonstrate parental care in dinosaurs it's necessary to find babies inside nests, an assumption I'd made after Bob Makela and I discovered the first maiasaur nest at the Willow Creek anticline. Among colonial dinosaurs, as among colonial birds, some young apparently left their nests early on, but if they were altricial they remained within the nesting grounds until they reached a certain level of maturity. This is why I said toward the end of the previous chapter that I wasn't surprised by the configuration of skeletal remains we discovered in the Milk River badlands. The fossil evidence—a heavy concentration of baby and juvenile bones inside the nesting horizon but few outside—indicated that the behavior of the unidentified lambeosaurs of the lowland Judith River Formation was consistent with that of the lambeosaurs of the upland Two Medicine Formation. Only under unusual circumstances, I now suspect, is it possible for a group of babies to be preserved as fossils within a nest. As is true of modern birds, the most common cause of death would have been parental abandonment. Those that didn't starve to death very likely would have been scavenged. Rarely in fact do solitary hatchlings or nestlings of any kind escape predation. After excavating several dinosaur colonies over a period of almost twenty years it has become clear that finding a nest containing babies is probably a once-in-a-lifetime experience, if it happens at all. At the Willow Creek anticline in 1978, Bob and I were a great deal luckier than we imagined.

Hypacrosaurus stebingeri has proved to be an ideal dinosaur to compare with *Maiasaura peeblesorum*, allowing us to test the ideas about parental care the latter had inspired. Both are large, plant-eating duckbills that lived along the upland reaches of the coastal plain during the late Cretaceous. Both traveled in large herds and nested in colonies. Also crucial, though not a characteristic per se, both left behind ample evidence of their existence in the fossil record. This is not to say there aren't differences between the two, of course. Perhaps the most notable concerns their offspring. Though the baby hypacrosaur skeletons at Blacktail Creek North vary in length, all are appreciably larger than the baby maiasaurs at the Willow Creek anticline.

A similar pattern emerged in 1988, when Philip Currie and a crew from the Royal Tyrrell Museum of Paleontology unearthed a third hypacrosaur colony in the upper layers of the Two Medicine Formation, supplementing the embryonic and nestling specimens recovered at Landslide Butte and the Two Medicine River, which already constituted the most extensive such collection of any single duck-billed dinosaur in the world. Located in extreme southern Alberta, only a few miles north of the Landslide Butte rookery and resting on the same sedimentary horizon, Devil's Coulee, as the site is called, has yielded several nests, a large number of eggs, and a broad assortment of embryo bones, most of which are intact. Six years after Phil's discovery, he and I published a paper in which we summarized what we had learned about the skeletal material preserved within the three nesting grounds. Near-term hypacrosaur embryos, it turns out, are about twenty-eight inches long, nearly the size of the fifteen baby maiasaurs in the nest from the Willow Creek anticline. Larger at birth, hypacrosaurs also were larger when they left their colonies, about four and a half feet in length, whereas maiasaurs apparently fledged after reaching three and a half feet.

As had happened when Bob Makela and I first published our conclusions about *Maiasaura,* the paper Phil and I wrote prompted a new round of challenges to the parental care hypothesis. Although the critics adopted different lines of attack, they all leveled the same charge: If the maiasaur skeletons that we were calling nestlings are only slightly larger than hypacrosaur embryos, then the maiasaur skeletons may well represent embryos as well, or at least very recent hatchlings. What we'd found at the Willow Creek anticline, in other words, were not nest-bound babies in need of adult care and protection but near-term or recently hatched youngsters that happened to die just as they were about to leave. One critic pointed to the handful of eggshell fragments present in the bottom of the nest, arguing that since altricial birds remove such material to make room for their offspring following birth, *Maiasaura* probably wasn't altricial. Others made much of a finding that, quite frankly, had puzzled us at first, too—the teeth of the hypacrosaur embryos, which are diamond-shaped when new, show distinct signs of wear. This was surprising because the teeth in the fifteen maiasaur skeletons are worn flat as well and I had taken that as evidence that they had been eating for some time, which is to say, they were nestlings, being fed by their parents.

Clearly I'd jumped to the wrong conclusion concerning the teeth, a mistake I'll correct shortly, but the primary charge—that the *Maiasaura* nestlings are actually embryos—is easy to put to rest. My first response is that the critics are overlooking a crucial piece of evidence: We have found dozens upon dozens of hypacrosaur and maiasaur eggs and they aren't the same size, not by a wide margin. To be able to say exactly how wide, I decided to reexamine the eggs, an easy procedure under normal circumstances, but at the time the *Maiasaura* clutch was part of a traveling exhibit touring the U.S. and wasn't scheduled to return to the museum for months. I eventually caught up with the tour at the South Carolina State Museum, where I happened to be giving a lecture, and while there I measured the eggs. Average diameter? Four inches. The average diameter of hypacrosaur eggs, by contrast, is nine inches. The difference in volume is especially telling, with maiasaur eggs averaging about 80 cubic inches, hypacrosaur eggs, 215 cubic inches, or almost three times the capacity. Picture a big grapefruit alongside a soccer ball. Using a graphic computer program that allows us to decrease or enlarge the size of three-dimensional figures and fold them into various positions, my wife Celeste and I have estimated that a near-term embryo could not have grown longer than eighteen inches within the confines of a typical maiasaur egg. As for the two-and-half-foot dinosaurs we found at the Willow Creek anticline? Only a Cretaceous magician could have produced animals of that size from hats only four inches wide; it surely wasn't within the power of *Maiasaura*.

Those convinced that dinosaurs were precocial have also suggested that the nesting grounds we've found—all of them, belonging to *Hypacrosaurus* and the unidentified lambeosaur as well as *Maiasaura*—best resemble contemporary crocodile rookeries. They note in particular that crocodile parents place their babies in a common area of water, an aquatic nursery, to which thereafter any adult might come to rescue a youngster in peril. If duck-billed dinosaurs followed a similar practice, it would explain the large numbers of small bones on the nesting horizons without relying on a parental care scenario. The flaw in this argument, however, is that nursery crocodiles are capable of capturing their own food and they can protect themselves from most if not all predators—they bite back. But newborn maiasaurs, only five or six inches tall, weighing perhaps two and a half pounds, and lacking sharp teeth or claws, were completely defenseless. To a hungry pack of theropods—troödons, say—the baby duckbills would have been about the the most vulnerable dinosaur prey on the coastal plain. And baby hypacrosaurs, though bigger than their maiasaur counterparts, were no better equipped to repel deadly carnivores. If groups of *Deinonychus* juveniles attacked one-and-a-half-ton tenontosaurs, as our excavation in the Cloverly Formation near Rainbow Butte seems to demonstrate, then groups of carnivorous theropods could probably subdue ten-pound hypacrosaurs, and with very little effort. Indeed, had such predators been persistent, the only effective deterrent would have been a herd of adult hypacrosaurs acting together to guard the perimeter of the nesting ground; in the process, some of them, though twelve feet high and thirty feet long, would have lost their lives.

In my opinion, one can make a strong case for parental care based on the size of eggs, embryos, hatchlings, and nestlings found on nesting horizons. But there's another survival strategy whose presence among hadrosaurs and lambeosaurs is overwhelmingly persuasive, even in the absence of any other proof. In studies conducted initially with David Weishampel and more recently with Celeste and graduate student Beverly Eschberger, we have determined that for a long while after birth neither *Maiasaura* nor *Hypacrosaurus* nor the Milk River lambeosaur could stand on its own, much less walk long distances to food sources outside the colony or, more difficult still, run from predators. Like altricial birds, their young skeletons were not sufficiently developed for such activity. The studies are worth recounting in some detail because the evidence they provide for parental care is more direct and more dramatic than any other. In addition, the particular methods we employed, borrowed from a field called histology, underscore the growing importance of the laboratory in dinosaur paleontology.

Here's a shorthand definition of histology: anatomy through a microscope. Here's a more elaborate version: the study of the internal structure and composition of plant or animal tissue at exceedingly fine levels of detail. At the museum laboratory, one of a handful in the world equipped for this kind of work, our particular interest is osteohistology, the microanatomy of bone, because, of course, that's all that remains of the dinosaurs. Actually, if I wanted to be precise, I'd have to say that what we practice is paleo-osteohistology, the detailed study of very old bones. Regardless of the age of the specimen, however, the procedure remains the same. To examine, say, a hypacrosaur femur, we must prepare a section that's thin enough to allow light to pass through it. Using a special circular saw, we first cut from the bone a slice roughly $\frac{1}{32}$ of an inch thick. The slice is glued to a wafer of glass, then very slowly and very carefully ground on a horizontal wheel until it's about as thin as tissue paper.

What does one see when one looks at a so-called thin section of a hypacrosaur femur through a powerful dissecting microscope? That depends on a number of factors—which part of the

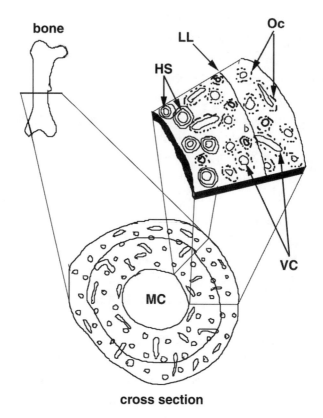

cross section

Cross section of a leg bone showing the internal microstructure.
VC = vascular canals; Oc = osteocytes; HS = Haversian system;
MC = marrow cavity; LL = lag line.

bone the sample represents, the age at death of the animal to which the sample belongs, whether the animal had been diseased or injured. What one sees also depends on what one knows about skeletal growth and structure. Reading bones, like reading sediments, is impossible without an understanding of the vocabulary bones use to "speak," as it were. For starters, you might be surprised to learn that the comparison of bones and rocks is not entirely metaphorical. Bones are formed in part by deposition and erosion of a mineral, calcium. Early in the embryonic stage, the skeletons of all higher vertebrates, including dinosaurs, are made entirely of cartilage, which, as the animal matures, is replaced by calcium and becomes, first, calcified cartilage, then honest-to-goodness bone, a process known as ossification.

The rate of growth and the pattern ossification follows leave characteristic traces in the microstructure of the skeleton, especially such long bones as the femur, which in large animals increase significantly in size between infancy and adulthood. Bone expands by means of two processes, one that widens it, called perichondral growth, and another that adds to its length, called endochondral growth. The faster a bone widens, for example, the greater are its needs for nutrients, and thus the blood vessels that convey the nutrients, and thus in turn the minute tunnels, called vascular canals, which contain the vessels. If in a thin section of the outer, perichondral layer of a femur you see lots and lots of canals, running along the length of the bone and extending sideways, or radially, like spokes, then you can be sure that the femur was well vascularized and growing rapidly at the time the animal died. By the same token, if few canals are evident and, moreover, there are distinct rings, called lag lines, indicating the cessation or near cessation of growth, you know that the femur was developing slowly, at least when the rings appeared. What's more, in bone deposited gradually, the tissues tend to be arranged in a linear, side-by-side pattern, whereas the tissues of rapidly growing bone, which are deposited more haphazardly, produce a characteristic woven matrix.

Later in the chapter I'll come back to perichondral growth as it relates to temperature regulation (the warm-blooded/cold-blooded controversy), but first let's look at the feature that's most relevant to the issue of parental care: endochondral growth. During the transformation that occurs in the skeletons of higher vertebrates, the ratios of cartilage and calcified cartilage to actual bone vary widely, depending on the maturity of the animal as well as on the developmental strategies it employs. Thanks to pioneering research conducted by the German morphologist J. M. Stark and others, we know that at hatching, for example, altricial birds possess two to ten times less ossified tissue than do precocial birds. We also know where the different tissues are located. In the legs—the femur and tibia—of altricial hatchlings, for instance, cartilage is concentrated at the ends of the bones, where they are attached to other bones. The point to remember about all of this is that there's an ironclad association between the skeletal anatomy of young birds and the behavior of their parents. If the adult form of a certain species takes care of its newly born offspring you can be sure that the leg bones of the offspring are only partially ossified at birth. By the same logic, if the leg bones are partially ossified, you can be sure that the off- spring require parental help to survive.

With this information in mind we compared the embryonic, hatchling, and nestling leg bones of *Maiasaura* and *Hypacrosaurus*, the embryonic and nestling leg bones of the Milk River lambeosaur, and the embryonic and early juvenile leg bones of the theropod *Troödon* and the ornithopod *Orodromeus*. In the case of *Maiasaura* and *Hypacrosaurus*, we used a computer graphics program to identify which skeletons in our collection best represent full-term embryos or hatchlings, a two-step process by which we identified the most effi- cient position an embryo could assume, then determined the largest possible skeleton that would fit in that position inside an average egg. Since we used only skeletons, making no attempt to reconstruct the bodies of the young dinosaurs, actual hatchlings were most likely smaller and less developed than those in our study. Moreover, only calcifed cartilage fossilizes, not cartilage itself, which means that what survives in dinosaur bones represents less of the nonossified tissue than was actually present in the animal when it died. Next, representative femurs and tibiae were selected for histological analysis. From the ends of each bone we removed at least three thin sections, cut lengthwise to expose the transition between ossified and calcified tissue.

The results were dramatic. We found that the near-term embry- onic leg bones of *Maiasaura, Hypacrosaurus,* and the Milk River lambeosaur are roughly 90 percent calcified cartilage, 4 to 7 percent bone, and 2 to 4 percent marrow. Among hatchling maiasaurs and hypacrosaurs the amount of calcified cartilage still represents 74 percent and 72 percent, respectively, of the total volume, while

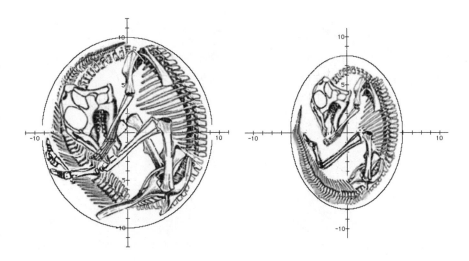

Hypothesized full-term embryos of *Maiasaura peeblesorum* (*right*) and *Hypacrosaurus stebingeri* (*left*) in reconstructions of their respective eggs. (*Skeletons based on* Hypacrosaurus *embryo by D. Sloan in Horner and Currie, 1994*)

about 10 percent is fully ossified, a sevenfold difference between the two kinds of tissue. In other words, more than two-thirds of the leg bones of baby duck-billed dinosaurs isn't bone at all but a material that's weak and brittle and thus poorly suited for the act of standing upright, still less for walking or running. In the single maiasaur nestling we examined, the amount of calcified cartilage dropped significantly, to 25 percent of the total, and the amount of bone increased to 20 percent, whereas the figures for the lambeosaur nestling are 50 percent and 17 percent.

Since we have a great variety of hypacrosaur specimens, we looked at samples from three different nestlings, each larger than the last. In the third and most mature femur the amount of calcified cartilage drops to 6 percent, bone rises to 30 percent, and marrow takes up 64 percent. Remember that immediately following hatching there was seven times more cartilage than bone. Now the com-

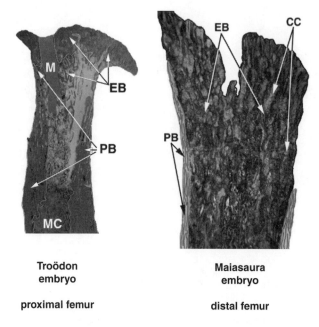

Troödon embryo

proximal femur

Maiasaura embryo

distal femur

Longitudinal sections of embryonic femurs of *Troödon* and *Maiasaura* showing the internal structure of the bones.
CC = calcified cartilage; EB = endochondral bone;
M = sediment; MC = marrow cavity; PB = perichondral bone.

position is almost reversed, with five times more bone than cartilage. By contrast, we found that the near-term embryonic leg bones of *Troödon* are made up of about equal parts calcified cartilage, bone, and marrow. And an even more striking difference is exhibited by *Orodromeus*, in which only 20 percent is calcified cartilage and 40 percent, or twice that amount, has already become bone. The implications, I believe, are clear. As in the case of altricial birds, the small amount of ossified tissue in the near-term embryos and hatchlings of *Maiasaura, Hypacrosaurus,* and the Milk River lambeosaur is the surest evidence short of direct observation that baby duck-billed dinosaurs were helpless. They couldn't have survived without assistance from adults. Indeed, like robin chicks, they couldn't have done much more than flop their heads back and forth. *Troödon* and *Orodromeus,* on the other hand, produced hatchlings in which the pattern of ossification more closely resembles that of alligators and precocial birds. In all likelihood, they could leave the nest immediately after birth.

In fairness I should add that other researchers have interpreted the same and similar skeletal remains very differently. Recently two graduate students from Oregon State University, Nicholas Geist and Terry Jones, under the guidance of physiologist John Ruben, examined the femur of a maiasaur hatchling, as well as the embryonic pelvises of *Maiasaura, Hypacrosaurus, Orodromeus,* and two other dinosaurs. After visually inspecting the bones, Geist and Jones became convinced that they are well ossified and on that basis concluded that the babies to which the bones belong were precocial. The flaw in this approach, however, is that ossified tissue cannot be accurately identified without conducting detailed histologic investigations, without, that is, preparing thin sections and studying them under powerful microscopes. When exposed to polarized light, true bone can be identified by the distinctive woven or linear patterns the tissue exhibits. In calcified cartilage, on the other hand, you see nothing but small circular structures, remnants of the pockets in which cartilage cells once were growing. None of this is visible to the unaided eye. Having mistaken calcification for ossification in both legs and pelvises, the critics were led to the erroneous conclusion that all hatchlings, including those of *Maiasaura* and *Hypacrosaurus,* were self-sufficient.

Besides strongly suggesting that the duck-billed dinosaurs were altricial and thus dependent on adult care during early development, the large amount of calcified cartilage in the leg bones of near-term embryos and hatchlings is indicative of rapid bone growth, as is the presence of extensive vascularization, another prominent feature of the microanatomy of baby skeletons. When cartilage cells mature they swell up, expanding the length and volume of the bone. The areas around the cells eventually calcify and the cells vacate the resulting shells, which in turn serve as scaffolding for ossified tissues. In the final stage of the process all of the original calcium is replaced by true bone. When there's a great number of swelling cartilage cells stacked in columns at the ends of bone, the bone expands at a fast rate. This is precisely what we see in *Maiasaura, Hypacrosaurus,* and the Milk River lambeosaur. And the two survival strategies—parental care and rapid growth among the young—were complementary, especially in the large dinosaurs.

Imagine: Maiasaur, hypacrosaur, and Milk River lambeosaur hatchlings were approximately ten times shorter than their parents. Rapid growth was absolutely necessary for survival, because the faster they matured, the less time they would have been vulnerable to predators and other mortal threats, including being accidentally trampled. *Maiasaura,* which weighed about two and a half pounds at birth, and *Hypacrosaurus,* about ten pounds, would have found it very difficult to stay out of the way of adults ranging between two and four tons—had they been able to roam wherever they pleased, that is. Based on the taphonomic evidence I've seen at the several colonies we discovered in the Two Medicine and Judith River Formations, my guess is that whereas the babies were not nest-bound, they were probably confined to nursery areas within the nesting grounds, much like the baby pelicans on the island rookeries of Lake Bowdoin.

The risks and benefits of combining the survival strategies of parental care and rapid development can be seen today among altricial birds, whose eggs and babies also tend to be smaller than those of their precocial counterparts. In their prolonged role as protectors and food suppliers the adults are more likely to encounter predators and other hazards, but that makes it possible for their offspring to channel all their energy into growth. The only thing required of the babies is that they lift their heads and open their mouths at the appropriate times. Apparently the same was true of baby altricial dinosaurs. In *Troödon* and *Orodromeus* a different trade-off took place. For one thing, the adult forms are much smaller than fully developed duckbills, so the babies didn't have to grow as much to reach maturity. At birth, for example, *Troödon* was about a foot and a half from head to tail, then grew to a maximum length of about eight feet. The offspring could afford a slower growth rate, but only if they were mobile from birth onward, permitting them to secure food for themselves.

Even so, *Orodromeus* babies on their own would have been highly vulnerable to predators. Adults probably protected them as well.

What about the worn teeth that we found first in the jaws of maiasaur nestlings, then in hypacrosaur embryos? It's now clear that dental wear is not evidence of parental care. My original interpretation turned out to be a misinterpretation. Yet there is a preponderance of taphonomic and histologic evidence that both *Maiasaura* and *Hypacrosaurus,* along with the Milk River lambeosaur, were altricial. That leaves only one explanation: while inside their eggs, the embryos ground their diamond-shaped teeth together, a survival strategy that equipped them with an effective—that is, flat—chewing surface by birth. The moment they hatched they were capable of eating. And eat they did. Eating, in fact, was their sole occupation.

Because the findings go to the heart of the parental care issue, it's worth repeating what the bones have allowed us to hypothesize so far: Baby duck-billed dinosaurs were incapable of standing on their own legs. They couldn't walk or run. They therefore couldn't have fed or protected themselves. During the first stages of development—until, we estimate, they had almost doubled in length—they couldn't have survived without the help of adult duck-billed dinosaurs, either their parents or, as suggested by the distribution of baby bones in the hypacrosaur nesting grounds at Landslide Butte and Blacktail Creek North, the colony as a whole.

But the revelations don't end there. Further histological research conducted at the museum laboratory has helped us construct a clearer picture of certain aspects of dinosaur metabolism, especially temperature regulation, a still poorly understood process that's implicated in all survival strategies, including, indirectly, parental care. I say indirectly because the presence of warm-bloodedness in itself doesn't mean that an animal is altricial. Many birds, after all, are precocial, as are most mammals. But there are behaviors found only in altricial animals—brooding is a primary example—that would confer no evolutionary advantage in the absence of warm-bloodedness. In other words, the hen wouldn't sit on her nest hour after hour, day in and day out, if she weren't incubating her eggs with the heat she generates internally.

If you've heard anything about the warm-blooded–cold-blooded debate, you probably know that it's caused a great deal of controversy among dinosaur paleontolgists and other life scientists. It's also caused confusion among nonscientists trying to grasp why the scientists are making such a fuss over temperature regulation. Reading some popular accounts, for instance, one could get the impression that every animal on Earth, those that existed in the past as well as those alive today, falls into one category or the other, either wholly warm-blooded or wholly cold-blooded, and never the twain shall meet. But like everything else in the natural world, the actual situation isn't that simple. And if it isn't that simple today, there is no reason to believe that it was any simpler 70 or 100 or 150 million years ago.

Let's start correcting the oversimplification by introducing a more precise language for describing temperature regulation. We need to enlarge our vocabulary, because the terms warm-blooded and cold-blooded obscure the wide range of ways in which the metabolisms of different animals operate. Imagine trying to describe a rainbow using only the words black and white. It can't be done—not, at any rate, in a manner that will do justice to the richness of the phenomenon. A palette of words is needed, one that reflects all existing colors as well as the many ways that colors can be mixed and shaded. To describe the rainbow of temperature regulation, at least two crucial differences must be recognized at the outset. The first distinguishes animals that possess an internal regulatory system—a set of metabolic processes—that maintains body temperature from those whose body temperature depends on the temperature of their environments. Typical living endotherms ("inside heat") include birds and mammals. All other creatures—reptiles, amphibians, fishes, insects—are ectotherms ("outside heat"). This is not to say, however, that ectotherms can't regulate their own temperature, only that eating alone isn't sufficient. To get warmer they must amble into the sunlight.

The second distinction concerns internal temperature, whether it remains constant. Because all ectotherms depend to one degree or another on the environment, their internal temperature is subject to change, sometimes very dramatic change. For this reason they are known as poikilotherms ("varying heat"). Frogs and lizards are ectothermic poikilotherms. But some endotherms are poikilothermic as well—certain birds, for instance. Though they produce heat internally they don't maintain a constant high body temperature. In the hummingbird the difference between flying and resting temperatures is twenty degrees. When a turkey vulture can't find food it will slip into a state of torpor, a strategy by which its body temperature drops by as much as thirty degrees, permitting it to survive while expending a significantly smaller amount of energy. The hummingbird and the turkey vulture differ from, say, the chameleon in that, despite the variation in body temperature, they still generate heat internally. Perhaps, then, they should be called heterotherms ("different heat")—endothermic heterotherms—to distinguish them from true poikilotherms.

Don't fret if you find yourself getting a little disoriented at this point. Go back and reread the last two paragraphs. The payoff for absorbing what I just said is this: We tend to think of birds and mammals as warm-blooded creatures and leave it at that, ignoring the differences between the two groups. But mammals are unlike birds in that they are homeothermic ("same heat"). Not only do they produce heat internally, they maintain a constant high body temperature—an incredible 98.6 degrees in human beings—under all environmental conditions and during all activities. There's a great deal more at stake here than the accuracy of the nomenclature we use to talk about temperature regulation in animals. The reason lumping together birds and mammals is dangerous isn't that it obscures their differences per se but instead that it makes it all but impossible to appreciate what those differences imply about the natural history of the two groups. That avian temperature regulation differs from mammalian temperature regulation is no accident; the underlying metabolic processes didn't evolve at the same time or under the same conditions.

We can't be entirely sure, of course, but the evidence suggests that the earliest mammals were nocturnal. For one thing, their skulls have openings precisely where in modern mammals whiskers are located, and whiskers are sensory organs that animals use to find their way in the dark. For another, unlike birds and other creatures, most mammals are drab-colored and many of them are color-blind. Emerging during the Jurassic period, precisely when dinosaurs dominated all terrestrial environments, they compensated for their small size and relative vulnerability by evolving a survival strategy that allowed them to be active after dark, when temperatures fell. They developed a thermal engine to keep them at a constant high temperature no matter how cold it got outside. This was an expensive strategy, because it required the continual consumption of large amounts of nutrients, but it worked. Birds, on the other hand, which have always been diurnal, may have acquired their type of warm-bloodedness to cope with the cooler climate that became prevalent during the Eocene period, some 40 million years ago.

By recognizing this distinction we come to a deeper understanding of the complex historical relationship between birds and mammals. And reconstructing such relationships—identifying the features animals do and do not have in common, by virtue of their shared and separate life stories—is the only way to understand evolutionary processes, including those responsible for the appearance and disappearance of certain horned dinosaurs along the coastal plain during the late Cretaceous.

At the end of chapter 6, I described the correspondence between these evolutionary developments among ceratopsians and the ebb and flow of the Western Interior Seaway as an epic waltz in which the pattern the participants followed was clear but not the music that inspired it. Creation, in other words, was unfolding before our eyes while the mechanisms driving it remained hidden from view. Mapping defining characteristics (temperature regulation is only one of many) as they emerge, change, and diversify from one generation to another brings the mechanisms into focus. You don't know a dinosaur—or, for that matter, any other extinct animal—until you've met its entire family, ancestors and descendants, all of its kin.

9
SUSPICIOUS
SPECIES

Strangers meet and introduce themselves. If they are adults, two questions invariably arise: "What is your occupation? Who are you related to?" As society grows increasingly mobile and urban, it's true that the second inquiry tends to become less personal—"Who do you know?"—but the intent is the same. We orient ourselves with respect to others in terms of jobs and social connections.

Paleontologists, though certainly more rigorous, adopt a similar approach when they first encounter a new dinosaur. Essentially our research falls under two headings: what the creature did for a living, in other words, the strategies it used to survive, and its hereditary ties, where exactly it fits in dinosaur society and dinosaur history. For us, however, the two issues are inseparable, or should be; they are merely opposite sides of the same coin, because the legacy that one generation passes on to another consists of survival strategies, both behavioral and morphological. Thus, by switching from my original interest, social behavior, as I'm now doing, to large-scale patterns of speciation and extinction—or, the shorthand version, from eggs to evolution—I'm not really turning away from that interest so much as turning it over to inspect the other side.

Strictly speaking, I started the switch in the closing paragraphs of the last chapter, when I transformed the problem of how dinosaurs might have regulated body temperature into an inquiry into what the different modes of temperature regulation reveal about evolutionary relationships among all animals but especially those involving dinosaurs—who descended from whom, how it happened, when it happened. Just as during fieldwork paleontologists carry a search image—a set of expectations regarding where specimens will be found and how they will appear when they are found, as well as mental pictures of the environmental settings dinosaurs inhabited—during the analysis of specimens they make assumptions about taxonomy (the classification of animals), usually some mixture of what's generally agreed upon at the time and speculation derived from individual experience.

To take an obvious case, most paleontologists today—not all, but most—believe that dinosaurs gave rise to birds. Taxonomic assumptions (who descended from whom, how and when it happened) are extremely critical because they shape the way we interpret fresh evidence—evidence regarding survival strategies, the other side of the coin. Without some sort of global taxonomic scheme in mind, we would have no means for orienting ourselves with respect to this or that aspect of dinosaur life, no way of knowing where it belongs in the total evolutionary picture. By the same token, however, in our eagerness to orient ourselves we run the risk of misinterpreting or ignoring altogether evidence that doesn't fit into the accepted taxonomic scheme. A couple of examples should clarify this dilemma's significance.

As I explained in the preceding chapter, histologic studies conducted at our laboratory at the Museum of the Rockies revealed both extensive vascularization and large amounts of calcified cartilage in the long bones of baby maiasaurs, hypacrosaurs, and Milk River lambeosaurs. These features are consistent with rapid growth. Indeed, baby duckbill bones resemble those of altricial birds, which develop extremely fast; this in turn strongly suggests endothermy—some kind of warm-bloodedness. Yet those same bones, upon reaching maturity, undergo a dramatic change in perichondral growth. There are far fewer canals. Lag lines are present. The tissues exhibit an orderly layered pattern. In other words, growth has slowed to a snail's pace, or, more precisely, an alligator's pace. The bones of a full-grown duck-billed dinosaur look a lot like the bones of a half-grown ectotherm, that is, a cold-blooded reptile.

The ability to shift by adulthood from a high-octane birdlike metabolism to one more reminiscent of reptiles conferred obvious advantages on maiasaurs, hypacrosaurs, and Milk River lambeosaurs. They developed rapidly when they most needed to, while small and weak and highly susceptible to predation, injury, and disease. Dinosaurs, however, like alligators but unlike birds and mammals, never ceased growing from the day they were born till the day they died. If their metabolism had continued at the same accelerated rate throughout their entire lives, these already oversize animals would have attained staggering proportions. In truth, that rate could not have been been sustained. Simply consuming enough food to maintain a high body temperature would have been a very difficult if not impossible task. Instead, as they approached maturity their metabolism shifted into a lower gear, decreasing body temperature, and thereafter they required less food. They probably also benefited from a phenomenon known as mass homeothermy, which is the tendency of large animals to retain heat simply by virtue of their bulk, which would have reduced the need for internal temperature regulation in adulthood.

These intriguing observations aside, the point I wish to get across is that there are no living mesotherms ("middle heat")—my term for animals that possess a bimodal metabolism—or anything remotely resembling one. There's no creature that starts life with the bones of a bird and ends it with the bones of a reptile. Another way of saying this, one that underscores the connection between inherited survival strategies and evolutionary processes, is that duck-billed dinosaurs don't fit comfortably into any of the taxonomic categories we use to classify contemporary forms of life. They defy categorization. And I think the difficulty lies less with our understanding of dinosaurs, which is woefully incomplete, than with the deficiencies of the classification system.

Here's a second illustration of the problems one can encounter trying to cage dinosaurs in categories that were built to hold very different kinds of animals. In 1993 the American Museum of Natural History mounted an expedition to the Gobi Desert. In a region southwest of the Flaming Cliffs, where some seventy years earlier Roy Chapman Andrews's crew had found the world's first reported fossilized dinosaur eggs, the museum's associate curator of vertebrate paleontology, Mark Norell, and expedition leader Michael Novacek made an outstanding discovery. They found a nine-foot-long late Cretaceous theropod, called *Oviraptor,* cousin to *Troödon* and *Deinonychus,* squatting over a clutch of *Oviraptor* eggs, at last twenty of them. It was an extremely unusual find. Unlike zoologists, paleontologists cannot observe behavior; they must deduce it from fossils. The taphonomic and histologic studies of *Maiasaura, Hypacrosaurus,* and the Milk River lambeosaur that I described earlier are typical of this inferential approach. The *Oviraptor* scenario is as close as one is likely to get to seeing dinosaur behavior—at the very least, some kind of parental attention—recorded directly in rock.

Less certain, of course, is the precise nature of the attention the *Oviraptor* is giving to its eggs. And that's where the taxonomic assumptions come into play. Geist and Jones, for example, who believe that all dinosaur hatchlings were well ossified and thus precocial, assert that the behavior of the *Oviraptor* best resembles that of crocodiles, snakes, and the like, which also tend to their eggs. Crocodiles do indeed build nestlike structures and cover their eggs with dirt and vegetation. They then rest their chins, or occasionally their bodies, on the nest. Some snakes, for their part, wrap themselves around their eggs. But these are superficial similarities. The crocodile doesn't place its head on the eggs but instead on the debris covering the eggs, and it does so to monitor the temperature of the debris, which is fermenting and thus producing heat. The actual incubation agent, then, is the pile of vegetation, whose temperature the crocodile maintains by adding or subtracting material as necessary during the ninety-day incubation period. The snake's behavior is even less relevant. It doesn't lie on top of its eggs but around them, acting like a barrier. The apparent aim of its attention is protection.

Nevertheless, someone could insist that the crocodilian comparison in particular is a reasonable interpretation, certainly as valid as any other. And it is. But only if you ignore what the American Museum researchers actually found in the Gobi Desert. The dinosaur-egg scenario is much more than a "close association," as Geist and Jones have characterized it. The *Oviraptor* sits directly atop its eggs, which are arranged in a compact circle. Most important, the dinosaur's arms are bent backward, over the clutch, in a protective or embracing fashion, exactly like the wings of a nesting chicken. Indeed, in every respect the position of the *Oviraptor* resembles that of brooding birds. Yet Geist, Jones, and like-minded scientists choose to compare the dinosaur with everything but birds, asserting that it is more like a crocodile than it is like anything else alive today. Why is that? Because to admit otherwise would be tantamount to saying that the *Oviraptor*'s behavior resembles that of a warm-blooded animal. The main reason for an animal to brood, not merely to maintain a "close association" with its eggs but actually to assume a brooding position, is to transfer heat to its eggs, heat that's generated internally. The critics' taxonomic scheme—in particular, that dinosaurs are reptilian, thus ectothermic—won't allow them to acknowledge that, in this instance at least, a dinosaur is acting like a bird.

I've always paid close attention to the Gobi Desert excavations because of the great number of eggs that have been unearthed there. But the *Oviraptor*-egg scenario is of special interest, not only because of its bearing on our ideas about parental care but because it led to a startling discovery about some of the eggs we recovered at the Willow Creek anticline. The first *Oviraptor* was found during the

1923 American Museum expedition to Mongolia. Like Norell and Novacek's specimen, it rested atop a clutch of eggs. Since Andrews and his colleagues had unearthed an abundance of *Protoceratops* skeletal material and little else in the Flaming Cliffs area, they naturally assumed that the eggs belonged to the primitive horned dinosaur. The overlying skeleton, however, was clearly that of a new theropod, a bipedal meat eater. What they had actually discovered, it seemed, was a dinosaur caught in the act of predation. Hence the name: *Oviraptor philoceratops* (the "egg plunderer" that "loves horned dinosaurs"). But in 1993 Norell and Novacek found an embryo in the same kind of egg and it turned out to be *Oviraptor*. In all likelihood the small theropod stole the eggs of its horned neighbors whenever the opportunity arose, but that wasn't what it was doing in the original Andrews specimen. It was tending to its own unhatched offspring. In fact, only by making the connection between embryo and egg, then egg and skeleton, were Norell and Novacek able to identify the occupants of their *Oviraptor* nest.*

Reading that the first embryo of a carnivorous dinosaur had been excavated was thrilling. It also convinced me that finding embryos inside eggs is the only foolproof method for identifying the dinosaur that laid the eggs. And that in turn got me to thinking about Egg Mountain and a beautiful specimen that graduate student David Varricchio and I had dug up near Jack's Birthday Site. In 1993 David found a partial *Troödon* skeleton resting on a clutch of eggs, which we immediately presumed belonged to *Orodromeus*. We reached that conclusion because they resembled all of the other *Orodromeus* eggs in our collection: they were roughly lozenge-shaped, about nine inches long, with a lightly striated surface, and they stood on end in the nest. And the reason we assumed that the original eggs were deposited by *Orodromeus* is that since 1979, when we began excavating Egg Mountain, we found almost nothing but *Orodromeus* skeletal material near the eggs. We hadn't

---

*Recently Philip Currie of the Royal Tyrrell Museum of Paleontology in Alberta and Zhi-Ming Dong of the Academy Sinica in Beijing reported a second *Oviraptor* lying on top of a nest of eggs, adding further support to the brooding interpretation.

thoroughly examined an embryo, however. It hadn't seemed necessary. Norell and Novacek's reinterpretation of the Gobi Desert *Oviraptor*-egg scenario made me think otherwise. Had we been right about David's *Troödon*? About Egg Mountain?

Carrie Ancell, one of the world's finest and fastest fossil preparators, provided the answer in the fall of 1996. Using dental drills and ever steady hands, she deftly removed the hard sediment surrounding one of the embryos. Now we knew with certainty what the *Orodromeus* eggs contained, and it wasn't little orodromiads but instead little troödons. Except for the specific dinosaurs involved, our interpretation of David's *Troödon* bore an uncanny resemblance to the situation in Mongolia. We assumed that our carnivorous theropod had been sitting atop the eggs of an herbivore because a great number of the same eggs had been found in an area covered with the herbivore's skeletal remains. The theropod, however, had laid the eggs. That means that Egg Mountain is really a *Troödon* nesting ground, which in turn means that *Troödon*, despite being precocial, nested in colonies, crucial new information about an otherwise puzzling dinosaur. But the most important outcome of this humbling episode, apart from being reminded yet again that paleontology is a science for revisionaries, is the additional evidence it provides for the parental care hypothesis. David's *Troödon* seems to have been brooding when it died.

Those who are reminded of crocodiles when contemplating the egg-sitting *Oviraptors* of outer Mongolia probably won't picture birds when they hear about the egg-sitting *Troödon* of west-central Montana. And the public will be left wondering which, if any, interpretation is right, forced to choose on the basis of such irrelevant factors as whose voice carries the farthest or whose views receive the greatest amount of media attention. It doesn't have to be this way, however. I believe that equipped with a firmer grasp of evolution, just about anyone could form their own well-considered judgments of new discoveries in dinosaur paleontology. I also believe that evolution is still so widely misunderstood, even among scientists, that just about everyone could benefit from a quick review of the underlying principles. Last, and I promise to step down from the pulpit as soon as I've said this, I believe that the full meaning of the principles will elude us as long as we adopt the traditional taxonomic approach—the universally accepted method of classification by which we divide animals into groups called birds, reptiles, and the like.

Discarding the taxonomic system we all learned in high school is not as reckless as it may seem. But lest you doubt that the way we classify animals as a whole is essential to our understanding of dinosaurs in particular, let's reinvoke the griffin, the bird-mammal hybrid that inhabited Greek and Roman lore for hundreds of years. Recall that that imaginary creature more closely resembles its probable inspiration—the protoceratopsian skeletons of the Gobi Desert—than the reconstructions of fossil remains that took place much more recently, during the nineteenth century, for instance, when in Europe the first iguanodon was assembled. Why were prehistoric Mongolian nomads, supposedly backward and superstitious, more adept at interpreting paleontological evidence than post-Enlightenment scientists?

I think it was because in all of their desert wanderings the nomads had never encountered anyone with the ambition of Carolus Linnaeus, the Swedish physician and naturalist who in the mid-1700s took upon himself the herculean task of classifying every living thing on Earth, employing the first version of the now familiar scheme that begins with the categories species, genus, and family and runs through the increasingly more inclusive groups known as order, class, phylum, and kingdom. If the nomads had viewed the natural world in terms of the Linnaean system, they would have felt compelled to place *Protoceratops* in one category or the other, bird or mammal. They would not have been so open-minded about what their eyes told them—that the creature they found in the Gobi Desert possessed features from both categories and so belonged to both, or neither.

The British paleontologist Richard Owen, by contrast, was

operating well within the Linnaean system when he analyzed the fossil remains of *Iguanodon* and *Megalosaurus*. By the mid-1800s everyone was. Judging from the appearance of these terrestrial leviathans, an untutored nomad might have called them something like bison-lizards. Owen knew better. Assuming that all vertebrates fit into one of the five recognized Linnaean classes, he determined that *Iguanodon* and *Megalosaurus* are reptiles, creating a special suborder called Dinosauria. And from that point onward they were seen as such. One of the more misguided early consequences of this historic turn of events, as I noted earlier, was the tendency on the part of paleontologists to break bones to make them conform to preconceived, that is, reptilian ideas of dinosaur anatomy, posture, and movement. With that in mind, try to imagine what someone of that era, more to the point, of that persuasion, would have thought of discoveries like those we made at the Willow Creek anticline, especially the maiasaur babies we found inside a nest. Such evidence would not have been merely puzzling; it would have been inconceivable, literally so preposterous, so contrary to what everyone assumed was true that it could not have been seen for what it was. Dinosaurs are reptiles and reptiles don't take care of their young. The thought would not have crossed their minds.

But that was a long time ago, you might counter. Hasn't paleontology freed itself of these outmoded and arbitrarily restrictive categorical imperatives? Yes and no. Since the 1960s, when Ostrom and others argued persuasively that dinosaurs not only resemble birds but are related to them as well, paleontologists have been more comfortable with evidence that defies traditional Linnaean categorization. I don't know anyone today, for example, who believes that dinosaurs are reptiles in exactly the same sense that modern reptiles are, especially with respect to physiology. Yet Linnaeus continues to haunt us, and it's because of something he didn't take into account when he developed his classification system, something he couldn't have taken into account—because it didn't exist at the time.

Linnaeus introduced his way of thinking about organisms in the

1750s, before the ideas of extinction or evolution became widely accepted. Darwin, remember, didn't publish *On the Origin of Species* until 1859. Consequently, taxonomy, the system we still use for classifying every animal that ever lived, is based solely on the organisms that are alive today. Even more troubling, as Darwin's influence spread no one bothered to check if the criteria used to establish the standard taxonomic groups reflect the actual evolutionary relationships among the organisms included within those groups. And as it turns out, they do not. Indeed, the Linnaean taxonomic system is altogether blind to natural history. It's based on the assumption that animals don't have a history—ancestors that might have differed from their progeny. So long as one is concerned only with describing living animals, the system works well. But since it doesn't provide an accurate picture of kinship and lineage, it can't be used to reconstruct the patterns of change on which everything from individual development to the emergence of new organisms is based. Surely it's the wrong tool for trying to understand such long-extinct creatures as dinosaurs.

The most obvious and relevant example of the failure of the Linnaean system to reflect evolutionary patterns is the class Reptilia. Though most of us have been taught otherwise, the animals we usually include in this category do not have a common ancestor. In other words, there's no single defining characteristic that all of them have inherited, and thus unites them, yet at the same time distinguishes them from animals in other taxonomic categories. Some so-called reptiles—*Dimetridon,* the sail-backed lizard, for example—share more inherited characteristics with mammals than thay do with members of their designated class. Most of the others more closely resemble birds. Thus, to continue to use the term in the way that it has always been used is to obscure actual evolutionary relationships.

This point will make more sense if you take a close look at another misleading feature of the Linnaean system. Implicit in the overall taxonomic hierarchy is the assumption that members of the same rank are equal, species to species or phylum to phylum, but not members of different ranks, for example, species to order or species to phylum. This would be valid if animals didn't have histories, if they didn't evolve from other animals. They do, of course. And all of the seemingly reasonable categories in the Linnaean system start to show cracks when one travels back in time, tracing a particular animal's roots, until finally the entire system crumbles under its own weight. Remember when you were taught that fishes gave rise to amphibians and amphibians to reptiles? Assuming that that's in fact the case, was the first amphibian a species, family, or class? Can there ever be just one member of a species, family, or class? If not, how could a wholly new group of animals—the class known as amphibians, for instance—appear fullblown, all at once? And how about that ever-controversial idea that human beings descended from other primates? Creationists will always take exception to this claim, of course, but I have yet to meet anyone who can identify the point in mammalian evolution when apes turned into men.

Now we are getting to the crux of the problem. Truth be told, the entire preceding line of questioning was miscast. Evolution, in the strictest and most practical sense of the word, doesn't occur at the level of groups. Amphibians do not reproduce. Nor do reptiles, mammals, or primates. Evolution is instead the work of individuals, untold numbers of them, mating with each other and producing individual offspring that in turn mate with the individual offspring of other individuals. One animal at a time, one generation after another. This is why Linnaean taxonomy is the wrong tool for examining the natural history of dinosaurs. It obscures the very processes one is trying to see. Like a cheap microscope, it lacks sufficient resolving power. Even the lens known as species is incapable of bringing evolution into focus. Indeed, because the species concept is the fundamental organizing principle of the Linnaean system, it is more responsible than any other aspect of the system for the widespread confusion that currently exists regarding evolutionary processes.

*   *   *

What on earth is a species? To Linnaeus's way of thinking, it's a group of organisms that share a common essence, the same spirit, so to speak, which was consistent with his belief that taxonomy is the study of the underlying design of nature, the permanent, unchanging template for the overall arrangement of life and the shapes all of its manifestations assume. Later Darwin famously used the term in the title of his book. But in the course of replacing Linnaeus's static, idealistic view with a dynamic and historical one, he didn't describe a universal method by which one might always identify a species. And that's more or less how the situation stood until the middle of this century, when the American zoologist Ernst Mayr went at the problem from a different direction. Rather than rely on observable characteristics like anatomy and physiology, he proposed an operational definition: Any two organisms capable of begetting viable, fertile offspring belong to the same species.

Mayr's innovation was brilliant. In one simple stroke he eliminated the ambiguity and equivocation that had always attended efforts to define species in the traditional manner while simultaneously emphasizing that evolution occurs by means of reproduction. The single shortcoming, and for paleontologists it's a fatal one, is that the definition can be applied only to animals whose sexual behavior, along with the outcome of that behavior, can be observed—in other words, the living. There's no way to determine whether any two extinct organisms were capable of reproducing. Even if we find a pair of apparently identical fossils—say, tyrannosaur skeletons—lying side by side, we cannot, using Mayr's approach, tell if they belong to the same species. We are forced to rely on other, more circumstantial criteria, and for the most part those other criteria have been derived from traditional Linnaean taxonomy.

As a result, the concept of species has continued to be overemphasized in paleontology, skewing our pictures of the hereditary relationships among extinct animals (their family ties) and the manner in which their survival strategies (what they did for a living) emerged, changed, diversified, disappeared. Most misleading of all is the belief that evolution occurs on the species level, that natural history is the process by which one species is transformed into another. This belief is so deeply entrenched and persistent that it's led some life scientists to virtually the same conclusion as the creationists—that evolution can't be detected in the fossil record.

Here's how the argument goes: If you look at the entire history of life on Earth, it's obvious that change has occurred through time and, further, that certain trends have developed at the "higher" taxonomic levels. There appears to be an overall increase in complexity, for example, between the phyla sponges and vertebrates. But if you look at particular sections of the fossil record itself, what you see are shapshots of very different organisms separated by long periods of time. To account for these apparent "jumps" in evolution, the paleontologists Niles Eldredge and Stephen Jay Gould have proposed a theory they call punctuated equilibrium. Their basic idea is that speciation occurs in bursts, and when it does occur it happens so rapidly that it doesn't register geologically, that is, leave a clear trail of fossils in sedimentary rock.

But like the nineteenth-century Europeans who classified dinosaurs as reptiles, the punctuated equilibrium theorists are using a concept, in this instance species, that hides the very thing they are trying to observe. When they find an individual organism in the fossil record they call it a species. When they find another organism, very different from the first but obviously related, they call it a different species. Then, assuming that ancestor-descendant relationships occur at the species level, they are forced to postulate an undetectable speciation "event" to account for the change. Those who practice the dramatic arts have a name for such an event. They call it a deus ex machina, god from a machine. The term dates back to the early days of Greek and Roman drama when this or that deity would arrive onstage by means of a crane to dictate the outcome of a play. Now it refers to any agent or event that's introduced without warning to provide a contrived solution to an otherwise irresolvable situation. Perhaps it's unfair to imply that undetectable speciation events are contrived solutions to the otherwise insurmountable difficulties caused by the concept of species. But this much is surely true: There's nothing to be gained by debating their existence, since they are by definition undetectable.

No wonder so many nonscientists not only fail to grasp evolution but continue to doubt that it exists. As a paleontologist who enjoys sharing his work with the public, who believes that a publicly funded scientist should try to do so, this worries me, especially now that my interest has expanded from dinosaur social behavior to include dinosaur evolutionary processes as well. But there's more to it than that. As I've repeated frequently throughout this book, it's impossible to understand dinosaurs out of context. Initially I stressed the particular environments in which certain dinosaurs lived and died, as well as the other organisms, plants and animals, (including other dinosaurs), with which they shared those environments—dinosaur ecology, in other words. More recently, I've been focusing on the developmental and historical backdrop, the complex web of relationships that tie dinosaurs to other creatures before and after them. But the principle is the same throughout. A dinosaur out of context is like a character without a story. Worse than that, the character suffers from amnesia.

Let's go back to the basics of hereditary relationships and find the point at which everyone tends to go off track. Forget for the time being the theory of punctuated equilibrium and consider instead the more familiar, down-home criticism that creationists level against evolution. "You can breed dogs till you're blue in the face," they say, "but in the end you will always produce some kind of dog." Or, "No matter how often trout mate and bear young they will never mate and bear a frog."

My response to these assertions? They're right. Within limits, that is. When I said earlier that no one can pinpoint when apes turned into human beings, I was making a similar observation, though with a very different purpose in mind. The fact is, apes did not become human beings and fish will not, and did not, turn into amphibians. The problem again is one of resolution; when we use such terms as ape, man, dog, and frog we are focusing on the wrong level, wrong, at any rate, if what we are trying to observe is evolution in action. Consider again what we mean by evolution—change through time. It's that simple. And where does change take place? Certainly not at the level of a species, which is an abstraction, a word we use when referring to a group of organisms that appear to have certain characteristics in common. No, the change that interests us occurs among individuals, one generation at a time. It occurs exceedingly slowly, to be sure, so slowly in many cases that it's impossible for humans to witness firsthand. But change does occur. And once one realizes that, and knows where to look for it, it becomes obvious that evolution is taking place everywhere, all of the time. We are aswim in evolution at this very moment.

Consider the breeding of domestic animals. Both the Chihuahua and the Great Dane are descendants of the same ancestor, a wolflike creature in all likelihood. That's evolution at work. The son differs from the father who in turn differs from the grandfather. That's evolution, too. Whether the Chihuahua or the Great Dane differ enough from their common ancestor to call either of them a new species is simply a matter of definition. The point is that the changes that led to the emergence of the Chihuahua and the Great Dane took place on the level of the individual organism. The same is true of early primates and humans. The former didn't change into the latter; change occurred from one individual to another. Only over a very long period of time—millions of years—does it appear that species, or groups, change. But that apparent change is an artifact of resolution. As long as attention is focused on the species level, evolution is impossible to observe. And as long as we use the Linnaean system, that's where the focus will remain, which is unfortunate. The way we now categorize organisms breeds skepticism toward evolution, when in fact it's the notion of species itself that's most suspect.

Even when they know where to look, paleontologists face an impediment to observing evolutionary processes that other life scientists do not: In the case of extinct plants and animals the level of resolution is determined by the degree of completeness of the fossil record. As I discussed earlier, it is far from complete, very far. Indeed, we can be confident that almost all of the organisms that ever lived on Earth died without leaving a single trace of their existence. What's more, the traces that have survived are biased in favor of certain organisms, for example, the dinosaurs that lived along the upland coastal plain in Montana during the late Cretaceous. Those that inhabited the lowland region at the same time were less likely to be preserved, because their depositional environments were more caustic and less stable. And mountain dinosaurs from that period, if any existed, long ago went the way of their mountain habitats. Their depositional environments didn't survive ensuing geological processes in a recognizable form, so any organic remains they may have contained were erased as well.

There's only one way to compensate for the partiality and selectivity of the fossil record—by collecting as many and as various an assemblage of specimens as possible, bearing in mind that if evolution, change through time, is going to show up in the record it's going to do so as a sequence of similarities and differences among individual animals related by descent. Precisely which similarity or difference is passed on to succeeding generations is of course governed by the process of natural selection. The sauropod with the twenty-foot neck survives a drought because it can reach leaves its brothers and sisters, with necks only fifteen feet long, cannot. It alone reproduces, increasing the number of sauropods with twenty-foot necks, which in turn have their own long-necked offspring, and so on. Because all such characteristics are inherited, closely related dinosaurs share characteristics that other, more distantly related dinosaurs do not. Observing evolution, then, is a matter of identifying unique sets of shared characteristics that are present in one group and lacking in others. In practice, this means searching for features that show the degree to which one organism resembles or is different from another, and classifying them accordingly.

The procedure I've just described has a name: cladistic analysis. Developed some forty years ago by the German entomologist Willi

Hennig, cladistics has only recently been applied to dinosaurs; while the jargon associated with the procedure can be somewhat forbidding, it is a straightforward and very powerful way to bring into focus precisely that which the Linnaean taxonomical approach hides from view: evolutionary relationships. I'll try to keep the odd terms to a minimum but there are a few of them that can't be avoided.

The first two are derived character and primitive character, which refer to morphological traits that can be passed on from one generation to another. I'm introducing them together because they are relative terms. For example, one trait of particular interest to dinosaur paleontologists is the hip-socket hole, which, by allowing the leg to extend straight downward, enables an animal to stand upright. The hip sockets of all dinosaurs have holes in them, as do those of all birds. But in no other vertebrate is this the case. With respect to vertebrates overall we say that the hip-socket hole is a derived character, because it came late in the sequence of anatomical changes that occurred among the vertebrates. With respect to other Dinosauria, however, including birds, the hip-socket hole is a primitive character, because it appeared in the earliest members of that group of vertebrates. Among the Dinosauria overall, the hip-socket hole is called a shared derived character, because it's a trait that all dinosaurs and birds possess by virtue of having descended from a common ancestor. The aim of cladistic analysis is to identify clusters of shared derived characters that will enable us to determine how closely certain animals are related.

Despite appearances, the cladistic system of classification is refreshingly simple by comparison with the Linnaean approach. For one thing, the terms species, class, phylum, and the like are eliminated. There are no hierarchical rankings, only groups, called taxa, defined by a number of shared derived characters. For another, instead of conventional phylogenetic trees (reptiles came from amphibians, which came from fishes, and so on), which often confuse matters by grouping animals together simply because of certain superficial resemblances or because they happen to have lived at the

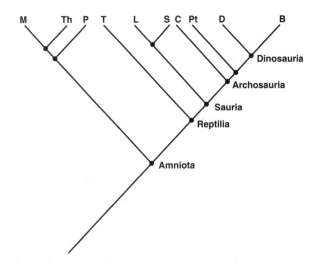

Cladogram showing the relationship of the amniotes. All amniotes lay eggs or give birth on land, having an egg with an amnion, and include the most recent common ancestor of living mammals and reptiles, and all its descendants. The Reptilia include the most recent common ancestor of turtles and the Sauria (including crocodiles, pterosaurs, dinosaurs, and birds) and all its descendants. The Dinosauria includes the most recent common ancestor of both dinosaurs and birds, and all its descendants. A cladogram shows evolutionary relationships.

same time, cladistic analysis produces cladograms that arrange clusters of taxa in a pattern that allows one to see evolutionary relationships.

That's enough new technical information for now. In the next chapter I'll demonstrate the usefulness of these ideas by conducting a simple cladistic analysis of the dinosaurs that roamed Montana's coastal plain during the Cretaceous period. I guarantee that after you have heard the evolutionary implications of our research of the past twenty years you'll have a deeper, more encompassing appreciation of dinosaur lives. But before moving on to that particular application, which I consider the climax of the book, I'd like to try to convey the overall conceptual flavor of my approach to evolution, which includes cladistics but isn't limited to it, showing that as a perspective on life it can help one sort through the claims and counterclaims surrounding temperature regulation, possible brooding scenarios, and other controversial issues in dinosaur paleontology.

Recall, for instance, that the punctuated equilibrium theorists propose sudden inexplicable pulses in evolution to explain the changes in organisms they observe when they look at life as a whole but which seem to disappear when they shift their attention to the species level. What they are saying is that they can find no transitional taxa, the so-called missing links, in the fossil record. What's missing, in other words, are actual evolutionary events. But in my view all taxa are transitional taxa. Life itself is transition embodied. Every living thing is in transition, a work in progress, every descendant a version of its ancestors—not an exact copy, but a version. The differences might be too slight to detect. Then again, they might not. Are there any noticeable differences between you and your parents? You and your children? And how about that weird uncle whose name hasn't been mentioned at the dinner table for the past ten years? When you flip through the pages of the family album you're witnessing evolution at work.

Are taxa missing from the fossil record? Sure. What's not missing, however, is transition. Everywhere you look you find transition. The holes in the fossil record don't represent mysterious speciation events. They merely mean that some taxa either weren't fossilized or haven't yet been found. I can't do anything about the former. So I go digging. Rather than arguing about the taxonomic category dinosaurs belong to, an activity that calls to mind a debate among librarians over the proper location of certain books, I search for more and more taxa, and more and more shared derived characters, then analyze the degree of relatedness among the taxa in terms of the characters.

And though I continue to employ words like mammal, bird, and dinosaur, I do so in the understanding that they don't refer to any specific organism but instead are abstractions whose only purpose is convenience, enabling me to talk about groups of organisms that have certain derived characters in common, yet differ among themselves in a million other respects. To think otherwise, to assume that there actually exists something that corresponds directly and wholly to the word mammal, is to engage in what I call the taxonomic fallacy. If we truly wanted to remain semantically faithful to the natural world, in which the only reality is the individual, we would assign a unique name to every organism that ever lived on Earth. That's impractical, of course. Instead we generalize. But as the saying goes, the map is not the territory. Words like mammal, bird, and dinosaur are simply conceptual tools—think of them as road signs—we use to orient ourselves as we travel through the welter of organisms that inhabit the wilds of evolution.

The reason I include cladistics in my conceptual toolbox is that, unlike the Linnaean system, it can be used to test hypotheses about evolutionary relationships. In setting up a cladogram, we identify and arrange clusters of shared derived characters. In the case of living animals we can use genetic, chemical, behavioral, and other characters, but the only indisputable characters that extinct animals exhibit are morphological, and those are further restricted to the skeleton—what the bones say about the life and times of their owners. (I think a few behavioral characters can be included in dinosaur cladograms, but this view is not shared by strict cladists.) Some characters are more diagnostic or indicative than others, of course, depending on the circumstances. Like all vertebrates, all dinosaurs have backbones, for example, but that tells you nothing about the relationship between hypacrosaurs and styracosaurs. I'll go into this in more detail in the next chapter, when I describe the duckbills and ceratopsians we found in the Two Medicine Formation. What I'd like you to keep in mind now is that if a group has been identified in terms of a cluster of shared derived characters, the fact that they also differ in some ways doesn't invalidate the grouping. The test isn't one of definition—which book belongs on which shelf, or the taxonomic fallacy—but of hereditary relatedness.

Here's what I mean: The reason paleontologists believe that the animals we call dinosaurs are closely related to the animals we call birds is that all of these animals have a large number of the same derived characters. The hip-socket hole, already mentioned, is one of them. It's found in no other vertebrate. Certainly the same character can evolve among unrelated groups, a situation cladists call parallelism. Eyes and wings, for instance, have appeared under different evolutionary circumstances. But it's exceedingly unlikely, verging on impossibility, for two unrelated groups to possess a large number of the same characters that other groups don't also share. Yes, dinosaurs and birds have a thin membrane that encloses their embryos, as well as backbones and differentiated heads, but so do mammals and reptiles. The diagnostic characters in this case are those that only dinosaurs and birds share, and the greater their number the more likely it is that the two groups are related. Besides the distinctive hip socket, these include the presence of five hip vertebrae; walking on three toes; a simple-hinged ankle; thighbones with an expanded ridge; and several others.

Nothing in paleontology is definitive, and cladograms are no exception. But if two taxa share a substantial set of derived characters seen in no other taxa, it's very likely that they also share a common ancestor. To be more specific, a thorough cladistic analysis of dinosaurs and birds strongly indicates that birds descended from a small, carnivorous dinosaur that was the ancestor of *Deinonychus* and *Velociraptor* as well. This being true, the fact that birds exhibit characters that dinosaurs don't or that dinosaurs exhibit characters also found in some reptiles doesn't mean that birds and dinosaurs are not closely related. Case in point: nasal turbinates. What is a nasal turbinate? A small structure in the nasal passage that captures moisture as an animal exhales so that it can be recycled back to the lungs when the animal inhales. Most mammals and birds have them, and if they didn't they would lose up to three-quarters of their daily supply of water. Mammals and birds, you see, breathe rapidly so as to consume enough oxygen to fuel their endothermic metabolisms. Nasal turbinates, therefore, are required for warm-bloodedness.

This, at any rate, is the position of John Ruben. The research behind it is his, too. And I think it's a very clever and fruitful line of reasoning. In fact, I loaned my CT scans of hypacrosaur skulls (similar to the scan depicted on the front of the book jacket) to Ruben's laboratory at Oregon State University so that he and his co-workers could look for nasal turbinates. But I don't agree that the apparent absence of turbinates in *Hypacrosaurus* and other dinosaurs Ruben analyzed necessarily means that dinosaurs didn't possess some form of warm-bloodedness. At this juncture, the turbinate research tells us that dinosaurs aren't exactly like living birds, and nothing more. It doesn't justify the conclusion that they are less like birds than they are like reptiles. One character cannot invalidate a relationship based on an entire set of derived characters that no other taxa share. It's entirely possible, for example, that nasal turbinates evolved after endothermy, as an anatomical structure that made accelerated metabolisms more efficient. And if dinosaurs were mesotherms, shifting at maturity to a slower rate, as the histologic evidence from duckbills suggests, they wouldn't have needed turbinates.

This debate is going to continue for a long time to come. My point is that cladistics, the reconstruction of evolutionary relationships based on groups of shared characters that could have arisen only by virtue of common ancestry, is the best tool for determining the implications of ambiguous or contentious evidence. Our judgments will change, certainly. They will undergo refinement. Sometimes they will be overthrown. An approach that incorporates cladistics won't bring certainty about anything. But unlike approaches based on the Linnaean system, it is self-correcting. Instead of forcing dinosaurs into preconceived taxonomic cages, we would do better by continually returning to the animals themselves, the individuals preserved in the fossil record, and always posing the same two questions: "What did you do for a living? Who were you related to?" If we're going to make progress in paleontology, that's the surest way to do so.

10
WITNESS TO CREATION

In his beautifully illustrated book *Serengeti,* wildlife photographer Mitsaki Iwago portrays immense herds of migrating wildebeests as they ford rivers or endure drought, often dying by the thousands, leaving behind carcasses that in turn make it possible for the region's scavengers, hyenas and vultures, to survive. The pictures serve as a vivid reminder that life and death are permanently entwined, a never-ending tango, sometimes painful, other times pleasurable, but in the absence of which no organism, neither you nor I, could exist.

Dinosaur paleontologists, it seems to me, are also scavengers of a sort. We labor in the graveyards of evolution, picking over bones, searching for clues to the lives of creatures that perished long ago. Until recently we were limited mostly to single specimens. On occasion, groups of up to twenty or thirty individuals have been found. And even less frequently, we've stumbled upon sites where the remains of hundreds of different kinds of dinosaurs have been washed together. But nothing like the Serengeti herds had been unearthed, preserved as fossils, until we discovered immense rook-eries and monospecific bone beds in sediments of the coastal plain that lay between the Rocky Mountains and the Western Interior Seaway during the Cretaceous period.

Hadrosaurs, lambeosaurs, and ceratopsians, in particular, but others as well, including small theropods and hypsilophodontids, evidently lived together—nested in colonies, migrated in herds, or hunted in packs. If nothing else, our excavations over the past twenty years have shown that many kinds of dinosaurs were much more social than previously thought. I've already talked at length about nesting, and described what little we know about hunting, but migration warrants a closer look. Years ago, when we had only the *Maiasaura* data to consider, I conjectured that the duck-billed dinosaurs migrated along an east-west route, laying eggs and raising their young in the uplands, then moving to the lowlands to live. I was led to that viewpoint by a bias in the fossil record—the appar-ent absence of eggs in lowland sediments. But by the early 1990s, when we excavated several nesting grounds in the Judith River Formation outside Havre, it had become clear that the herding dinosaurs, both horned dinosaurs and duckbills, were following an overall north-south trend.

In all likelihood the mass movements were driven by seasonal variations, just as the migrations of plant-eating animals are today. Notwithstanding the comparatively warm temperatures that per-sisted throughout the Cretaceous period, permitting crocodiles to survive as far north as Alberta, the angiosperms, or flowering plants, that had begun to flourish by that time would have lost their leaves periodically. Add to that the impact that hundreds, even thousands, of foraging dinosaurs, each weighing between two and four tons, would have had on the local vegetation and it's hard to imagine how the herds could have survived without moving regu-larly, if not constantly. To remain in one place would have caused mass starvation.

Precisely how far the herds migrated is uncertain, though there seems to have been some measure of variation. To date, large groups of maiasaurs have been found only in Montana, and only in the Two Medicine Formation, suggesting that that particular dinosaur confined its migratory travels to the upland plains. By contrast, the remains of *Edmontosaurus,* one of the largest duckbills, have been found in Wyoming, eastern Montana, Alberta, and the North Slope of Alaska, which is about as far north as one could have traveled then without crossing into Siberia. It's unlikely, of course, that the edmontosaurs of Alaska actually migrated to Wyoming, or vice versa. Distances of that magnitude are beyond the capacity of wingless creatures. But imagine, if you will, the time of the year when at extreme northern latitudes the sun scarcely rises above the horizon all day long. Would thousands of duck-billed dinosaurs, weighing five to six thousand pounds and ranging up to thirty-five feet in length, have remained there, milling about until the darkness lifted? I strongly doubt it. During that more temperate time there was no snow or ice to prevent them from moving. I think they would have followed the sun, migrating at least far enough southward to be able to see the vegetation they depended upon for survival.

Despite the uncertainty that remains regarding the direction and extent of migrations among duck-billed and horned dinosaurs, I have no doubt that they traveled in large social groups, very much like the bison of pre-Columbian North America and Africa's wildebeests, zebras, and kudus. That's the first thing the mass mortality sites tell us. They also tell us, as Ray Rogers determined in his convincing reconstructions of local environments in the Two Medicine Formation, that herding and migrating were survival strategies that also conferred disadvantages, all of which are variations on the following principle: Any threat to one was a threat to all. When drought or disease arrived, a flooded river had to be crossed, or violent volcanic eruptions occurred upwind in neighboring mountains, whole herds were placed at risk and consequently large numbers of animals perished at once.

And judging from the fossils found in the bone beds, death was indiscriminate. Among maiasaurs, for instance, we found skeletons that vary from nine feet to the maximum known length in adults—about twenty-five feet. (The fate of the intermediate juveniles, those large enough to leave their nesting grounds, which we estimate were about three and a half feet long, yet were smaller than the smallest members of the herds, is unknown. We haven't found their remains anywhere.) Besides duck-billed and horned dinosaurs, another group that seems to have coevolved with the angiosperms are the tyrannosaurs. The evidence is persuasive, I believe, that these large carnivorous theropods scavenged the carcasses of the social plant eaters whenever presented with the opportunity, which is to say whenever groups of the latter succumbed to the weather and other environmental hazards. In short, tyrannosaurs followed the herds, or at least remained in their vicinity, just as hyenas and vultures do on the Serengeti Plain today. Consider the scene: a mile-long parade of maiasaurs, thousands of them, crossing a stream made swollen and raging fast by relentless seasonal rains. Around the herd, at an inconspicuous but perceptible distance, stand tyrannosaurs and other scavengers, and they are aroused, nervous with anticipation. Quite an image, isn't it? To an insatiable old bone scavenger like me, it's even more arresting than Iwago's impressive photographs.

I'm now going to ask you to return with me to the graveyards of the Two Medicine Formation, not the nesting grounds but the sites where large numbers of young and adult dinosaurs died at the same time. I want to show you something that profoundly and permanently altered the way I think about the late Cretaceous ornithopods and ceratopsians of North America. Pick any one of these bone beds—the gigantic group of maiasaurs buried by volcanic ash at the Willow Creek anticline; the horned dinosaur herds that fell victim to drought and opportunistic disease near Landslide Butte, along with the groups of hypacrosaurs and prosaurolophs that died en masse in that area as well; or the gryposaur and maiasaur kill sites near the Two Medicine River. What do you see when you walk through the graveyards? Well, for starters, everything that we have already discussed—evidence of social behavior, geological clues to the types of environments the dinosaurs inhabited, indications of their cause of death. But have we missed anything?

Let's step back for a moment and reconsider what lies before us, to see if there isn't something else the graveyards might reveal, just as I did one day at a place we call Thunder Dome, a barren, bowl-shaped hill in the Landslide Butte badlands. If it's true that all the bone beds in the upper Two Medicine Formation represent herds of one size or another, then what exactly does that say about the behavior of the herd beyond the fact that they lived together, sharing resources? What else, in other words, do herds offer to its members besides protection? While pursuing that line of thought a realization dawned on me: If a group of similar animals travels together chances are it also interbreeds. In biology a group of interbreeding organisms is called a population. Moreover, it's considered the level at which speciation takes place. We're going to try to avoid that term, of course. The taxonomic name we use to refer to herds of maiasaurs, hypacrosaurs, and gryposaurs isn't important. What's important is that within the herd, individual animals were mating and producing offspring.

We can't be certain of this, surely. But I think it's a reasonable assumption. The bone beds contain the remains of interbreeding populations, or at least populations that were capable of doing so. As I sat on Thunder Dome thinking about the graveyards of the Two Medicine Formation it further occurred to me that while this isn't exactly what Ernst Mayr had in mind when he proposed his operational definition, it's as close an approximation as one can hope for with respect to fossils. It's certainly closer than I'd ever imagined getting in the study of dinosaurs. About the reproductive behavior of two or three or five extinct plant-eating animals there's nothing anyone can say that isn't pure speculation. The same is true of small groups of carnivores, the *Deinonychus* packs that preyed on tenontosaurs, for instance. But very large aggregates of plant eaters that exhibit all of the other characteristics of migrating herds? That's another situation altogether. Because the monospecific bone beds are populations in the full, biological sense of the term, they offer information that individual specimens and other kinds of aggregates don't.

The most important new information by far is the degree of individuation within the population. By that I mean the degree of morphological difference that thousands of animals might display and still be able to reproduce. To be specific, within the graveyards lay cross sections of the range of variation of crucial skeletal characters, for example, the length of ceratopsian horns and the shape of their neck shields. What is the value of such information? To illustrate, let's say that the horns on fully grown adults in one population range between thirty and thirty-six inches. If you find another ceratopsian outside the bone bed whose horns fall within the established limits for that character and that in every other important respect resembles the skeletons in the bone bed, then very probably it's a member of the same population. This is even more likely if the lone ceratopsian is recovered from the same sedimentary layer, which means that it lived at the same time. In short, knowing how much key characters vary within a population enables one to form hypotheses about the hereditary relatedness of different groups of dinosaurs.

Dinosaurs have never been subjected to this kind of analysis before. No one has made the attempt. In paleontology, population studies have been restricted almost exclusively to marine invertebrates, and for good reason: The fossils are small—indeed, tiny— by comparison with just about every kind of vertebrate, and thus relatively easy to excavate. What's more, there are tremendous numbers of them in the geologic record. Finally, the environments in which they were deposited—sea-bottom silt and mud—tend to be very stable, yielding sedimentary rock in which the evolutionary time frame is more clearly and completely preserved. Stephen Jay Gould, who has devoted a great deal of attention to ancient snails, among other seagoing creatures, is probably the best known student of extinct populations and surely one of the most provocative. His research will always be more quantitatively rigorous than mine. As dinosaurs go, the bone beds of the Two Medicine Formation are unusally populous, yet under the best circumstances, any conclusion I might reach about them will be measured in terms of tens of individuals. Invertebrate paleontologists, on the other hand, have access to thousands, even tens of thousands of the same kinds of specimens. Snail shells, I'm afraid, will always greatly outnumber ceratopsian skulls.

Even so, I think there's a decisive advantage to studying populations of dinosaurs. Like all other large vertebrates, ceratopsians retain characters whose functions are readily recognizable. Consider the horns and neck shields I spoke of earlier. Such prominent anatomical features are known collectively as sexual ornamentation, and their purpose, like that of elaborate elk antlers and brightly colored peacock plumage, is to attract prospective mates. The presence of horns on ceratopsians, then, provides an opportunity to observe sexual selection, the evolutionary process by which certain kinds of ornamentation change through time under the influence of environmental pressure. But what constitutes sexual ornamentation in snails? How do we determine which male features the female finds attractive? I'm not saying it's impossible to study sexual selection in snails but it's surely much harder than in dinosaurs. Indeed, morphological variation is so much easier to detect and measure in dinosaurs than in snails that any ideas we infer from them will be appreciably less equivocal and hence a good deal more relevant to our understanding of evolutionary processes. A sample that runs to ten thousand may give one statistical confidence in what one says about the sample, but if the specimens are so difficult to study that what can be said is inconsequential, what difference do the numbers make?

In the years since that modest epiphany on Thunder Dome, another dimension has been added, the last that we required to complete the reconstruction of dinosaur lives on the coastal plain. The work that Ray Rogers and I have done, characterizing environmental changes that took place during the late Cretaceous, especially events tied to the periodic advance and retreat of the Western

Interior Seaway, has provided a more detailed historical framework for understanding the dinosaurs that inhabited the coastal plain. Since the 1960s, certain geologists and paleontologists have speculated about correlations between the dynamics of the seaway and the evolution of terrestrial organisms living immediately west of the seaway (marine organisms were likely affected as well, and thus have been included in some of the speculations, but that's not our concern here). None of the proposed models for what transpired on the coastal plain during the late Cretaceous has been convincing, however, because no one had found terrestrial populations of sufficient size and distribution in the fossil record to support the models.

This changed when we discovered large dinosaur bone beds in the Two Medicine Formation. Not only did we have populations in which the evolutionary impact of environmental pressure could be observed, we had, thanks to Ray's work, populations whose positions in the overall historical context were clear. And position in this instance is of the utmost importance, because the bone beds represent crucial points during the application of environmental pressure along the coastal plain—namely, very near to a maximum point of regression, when the seaway had shrunk to its smallest size, and very near to a maximum point of transgression, when the seaway had increased to its largest size, flooding most of the land east of the Rockies.

The themes raised by these developments are the most difficult in all of paleontology, more than that, in the life sciences as a whole, and for that reason they often are avoided or taken up only by those of a theoretical turn of mind. These themes include the emergence, diversification, and disappearance of new characters and, ultimately, new kinds of organisms, as well as the ecological factors that affect such evolutionary changes. Even more uncommon is any attempt to specify the rate, direction, and mechanisms of evolutionary change, especially in the case of dinosaurs. Indeed, the bone beds of the Two Medicine Formation provide the first opportunity to do so. Viewed from the proper vantage point, those graveyards can be seen as a series of stages upon which the very processes of life itself become visible. Sitting atop Thunder Dome, we are granted the surpreme privilege of observing creation at work.

Thunder Dome, then, is where we finally gained enough resolving power to bring the evolutionary landscape into focus. This was made possible, it bears repeating, by the discovery of several large dinosaur bone beds that could be treated as populations, which in turn permitted us to observe transformational sequences, or change through time, at the level at which such change actually takes place—among individuals. But it wouldn't have happened without our also having developed a much more detailed map of the Two Medicine Formation, which enabled us to identify the precise locations of those populations in the geologic column.

Besides placing too much faith in the Linnaean taxonomic system, paleontologists have been too quick to embrace the system geologists use for classifying sedimentary rock. They sometimes forget that the development of that system was driven not by the search for fossils but by the search for fossil fuels. When petroleum geologists set out to describe the sedimentary rocks of Montana, they divided them into units, or formations, that made sense geologically. And as an aid to understanding the geological history and composition of the region, their maps and cross sections are very valuable—to the paleontologist as much as to anyone else. But as a method for categorizing organisms, geological formations can be misleading, in the same sense that the concept of species is misleading. They fail to resolve the geologic record to a fine enough scale to permit transformation sequences among fossils to become apparent. Unfortunately, some paleontologists continue to publish papers with titles like, "The Vertebrate Fauna of the Judith River Formation," implying that all of the animals catalogued in the paper lived at the same time, thereby further implying associations that do not exist while obscuring others that do. The Judith River Formation spans about five million years. A whole lot of evolution can go on in five million years.

The reason there's been so much speculation about the relationship between the Western Interior Seaway and terrestrial life along the Rocky Mountain Front, however, is that even when dinosaurs are lumped together by formation, the fauna that lived before the transgressions, especially the first and third, differ so markedly and with such consistency from those that lived after the associated regressions that it seems the rising and falling sea was somehow responsible for the shifts, a veritable engine of evolution. The pattern is simple, with the greatest degree of biological change following the most extensive environmental disruption, and that was the transgression represented today by the shales of the Colorado Group. (You might want to flip back to the diagram on page 47 as we take a second walk through this period of geological history.)

Picture once again the large-scale environmental events that took place on the coastal plain during the Cretaceous period. About 97 million years ago the seaway started to rise, gradually flooding the lowlands and, eventually, the uplands as well. The Colorado expansion continued for ten million years, at its greatest extent even invading the valleys of the newly forming Rockies. When the seaway stopped rising and began to recede, all of the land east of the Rocky Mountain Front was under water. There was no coastal plain 87 million years ago. All of the terrestrial habitat along the Rocky Mountain Front had been converted into aquatic habitat. The regression that followed occurred relatively quickly, over a period of about four million years. The second transgression, which started 83 million years ago and reached its zenith 79.6 million years ago, was comparatively small. But the third, which started 75.4 million years ago and reached its zenith about 74 million years ago, also significantly flooded the plain (unlike the first transgression, though, the third did not cover it entirely).

Let's also remind ourselves of the geological structures these events left behind. The terrestrial sediments deposited during the first transgression are known as the Cloverly Formation in the eastern part of the coastal plain and as the Kootenai Formation in the western part; the marine sediments are called the Colorado Shale. The terrestrial sediments deposited during the second transgression are known as the Eagle Sandstone in the eastern part of the coastal plain and the lower Two Medicine Formation in the western part; the marine sediments are called the Claggett Shale. The terrestrial sediments deposited during the third transgression are known as the Judith River Formation in the eastern part of the coastal plain and the upper Two Medicine Formation in the western part; the marine sediments are called the Bearpaw Shale. Above the Bearpaw Shale, consisting of terrestrial sediments deposited during the last regression of the seaway and afterward, are, to the east, the Hell Creek Formation and, to the west, the St. Mary River Formation.

Here's the pattern that, for some time now, has captured the imagination of paleontologists: In the Cloverly Formation we find two groups of ornithopods—tenontosaurs and hypsilophodontids; nodosaurs, which were a primitive type of ankylosaur, or armored dinosaur; the small theropod *Deinonychus* and the large theropod megalosaur; and various sauropods. These are the animals that lived on the coastal plain prior to the first and largest transgression. By the time the seaway receded, millions of years later, the cast changed substantially. Dinosaurs found in the Judith River Formation, which contains the remains of creatures living in the lowland plains prior to the third and final transgression, include hypsilophodontids as well as theropods—tyrannosaurs, for instance, and raptors.

But alongside these are entirely new kinds of dinosaur, pachycephalosaurs and their relatives the ceratopsians, as well as hadrosaurs and lambeosaurs. Where did the horned dinosaurs, in particular, come from? And how about the duckbills and crested duckbills? Obviously they are related to the tenontosaurs of the Cloverly Formation, but how exactly? The faunal changes that followed in the wake of the Bearpaw transgression weren't as striking but significant nonetheless. There are no traces of nodosaurs or lambeosaurs in the Hell Creek Formation, and new kinds of tyrannosaur, hadrosaur, and ceratopsia have appeared, once again raising questions about a possible relationship between the activity of the seaway and the fate of dinosaurs on the coastal plain.

It appears, then, that we have three distinct fossil assemblages that don't vary within themselves but differ greatly from each other. What's more, they are separated by varying periods of time, measured in the millions of years. Taken at face value, this is precisely the kind of pattern that inspired the punctuated equilibrium theory. The entire fossil record in fact looks the same—groups of very different organisms separated stratigraphically from each other, a record in which the intervals of silence are longer and more numerous than the intervals that offer information. In this instance, however, there's unmistakable evidence of environmental events of a scale and an intensity sufficient to account for the apparent jumps in evolution. Those who subscribe to Eldridge and Gould's perspective might venture to propose that the "speciation events" that presumably took place along the Rocky Mountain Front during the Cretaceous period, though undetectable, were governed by the dynamics of the Western Interior Seaway. But that's all they would venture. And even at that, at the level of resolution represented by the three fossil assemblages, the proposal is untestable. The fossil record doesn't provide enough of the right kinds of information.

I wasn't satisfied with our ideas about the evolution of dinosaurs on the coastal plain. They were little more than guesswork, guesswork of a sophisticated order, to be sure, and with which I was largely in agreement, but guesswork all the same. In particular, I was convinced that we didn't need to conjure up a deus ex machina to account for the three bursts of novelty in the fossil record. If we studied the terrestrial sediments in greater detail—increased the resolution, as it were—we would most likely uncover transformation sequences, that is, evidence of variation among populations within the faunal assemblages of each formation that would in turn reveal the manner in which the different populations are related. Even more important, we would at long last be able to hear the music that played during the evolutionary waltz on the coastal plain—in other words, actually identify the mechanisms that linked the lives of dinosaurs to the rhythms of the sea.

We couldn't search for fossils everywhere at once, of course, so the first task was to select the most promising sediments to explore. From a practical standpoint, the evolutionary changes associated with the Colorado transgression appear too difficult to study, at least for the time being. For one thing, the stratigraphic distance between the Cloverly and Judith River formations is so great, representing twelve to fifteen million years, that other factors beside the advance and retreat of the seaway may have contributed to the changes that occurred between the older and younger faunal assemblages. Determining whether that's the case and sorting through all of the possibilities would require an enormous amount of research. For another, apart from a few notable exceptions, the Cloverly Formation has yielded a small number of taxa fossils; the Kootenai Formation, very few; and the valley sediments, representing the area to which the animals that survived the flooding of the coastal plain would have retreated, no Cretaceous dinosaur remains at all (not in Montana or Alberta, at any rate). Likewise, the intervening lower Two Medicine Formation and Eagle Sandstone—which are, respectively, the upland and lowland terrestrial sediments deposited the first time the seaway retreated and the plain expanded—have produced very little relevant fossil material.

In rock deposited more recently, by contrast, dinosaur bones are plentiful. Most of the skeletal remains that Gilmore and Brown collected, for example, are from the upper Two Medicine Formation, and during the past twenty years I and my crews have recovered thousands upon thousands of specimens from the same sediments. Indeed, thanks to that work, which was motivated at first by a desire to find eggs and babies, and the fact that the formation as a whole is well preserved—some two thousand feet representing six to seven million years—in north-central Montana, I found myself in a position to examine evolutionary processes that corresponded with the Claggett regression, starting 79.6 million years ago and ending 75.4 million years ago, about four million years later, and the Bearpaw transgression, when the seaway

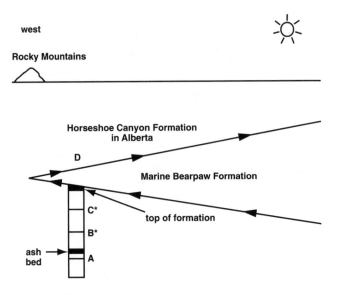

Cross section showing the top of the Two Medicine Formation, and the relationship of the four centrosaurine skulls. D is *Pachyrhinosaurus*. The arrows show the direction the shoreline was moving.

reversed course and began to rise again, reaching its maximum point of incursion onto the coastal plain 74 million years ago, about one and a half million years later. To be specific, our excavations at Landslide Butte and the Two Medicine River enabled me to resolve the upper Two Medicine Formation into several groups of fossils, each representing an upland population, some of which are separated from the others by various lengths of time.

As you may recall from the fieldwork chapters, the Landslide Butte sites include three new groups of horned dinosaur (from three stratigraphically discrete layers in and around Canyon Bone Bed, one of which is identical to the sediments of nearby Dino Ridge). To avoid the impression of taxonomic hierarchy, I'm going to refer to these new horned dinosaurs as centrosaurines, because that's the family to which they and their apparent relatives belong. Included among the other Landslide Butte sites are one each of *Hypacrosaurus stebingeri* (Lambeosite) and

Prosaurolophus blackfeetensis (Westside Quarry). The dinosaurs at the Two Medicine River sites are *Gryposaurus latidens* (Hillside Quarry) and *Maiasaura peeblesorum* (West Hadrosaur Bone Bed).

The three centrosaurine bone beds at Landslide Butte, dating to a several-thousand-year-long period about 74 million years ago, coincided with the final stage of the Bearpaw Transgression. Those bone beds therefore represent a chronological series of dinosaur populations being subjected to environmental pressure—significantly reduced land surface in the face of an advancing sea—at its most extreme. Although the exact point when the flooding stopped and the water began to fall again has not been located, it's been estimated that at that time the plain was only thirty to fifty miles wide, as little as one-eighth of its original size. Let's begin there, then, at Canyon Bone Bed and Dino Ridge, among the centrosaurines that, along with all of the other animals inhabiting the coastal plain about 74 million years ago, were being crowded into a progressively smaller habitat and thus, as time went by, competing for fewer and fewer resources.

Although back in chapters 4 and 5 I described the skeletal variation exhibited by the three different groups of horned dinosaur, I'm going to quickly recap that information now, since it is the foundation for my argument about the evolutionary consequences of the Bearpaw Transgression. Don't forget that each group represents a population that perished together, which allowed us to determine the maximum variation of any particular character within the population. Also keep in mind that we scoured the area over a period of five years, collecting thousands of specimens, without finding any evidence for overlap among the three populations. Each was recovered from a separate stratigraphic horizon, which is to say that they lived at different times. This gave us a basis for comparing characters that varied from one population to another through time. And in the case of the horned dinosaurs of Landslide Butte, the characters of interest are features of the skull, the horns and shields especially.

The most primitive of the three specimens has a long, forward-

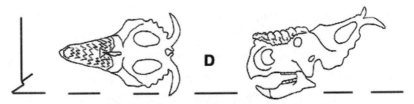

MARINE BEARPAW FORMATION

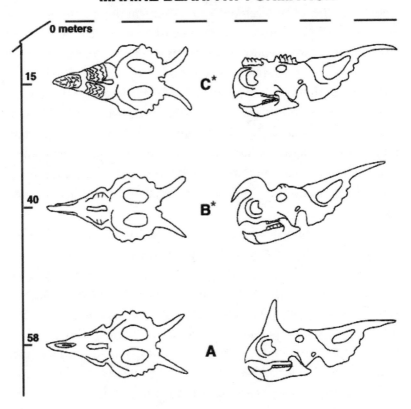

Centrosaurine skulls in the order in which they are found at the top of the Two Medicine Formation and bottom of the Horseshoe Canyon Formation, with the intervening marine Bearpaw Formation. The asterisks designate the transitional forms. The numbers show the distance the skeletons were found from the top of the Two Medicine Formation. (*Drawing based on an illustration originally by Kris Ellingsen, and additional morphological data.*)

pointing nose horn whose tip bends slightly backward. The areas above the eye sockets are slightly raised. Two long spikes extend from the neck shield, one on each side of the head. Significantly, this dinosaur bears a strong resemblance to *Styracosaurus,* from the older lowland sediments of the Judith River Formation of Alberta. Although *Styracosaurus* has three spikes on each side of its neck shield, the middle two are significantly longer than the others. The animal also possesses a long, forward-pointing nose horn and small bumplike ridges above its eye sockets. Although for the sake of convention, the first horned dinosaur of Landslide Butte soon will be given a traditional genus-species name, in keeping with my comments on the species concept, and to simplify this discussion, I'm going to refer to it as centrosaurine.1.

The second specimen in the series, known offically as *Einiosaurus procurvicornis* but herein called centrosaurine.2, is distinguished by a short nose horn that curves over the front of its face so that it is parallel with the snout. This is the intriguing hook-nosed horned dinosaur we discovered at the end of the 1986 season. Centrosaurine.2 also has a single spike on either side of its neck shield, along with knobs over its eyes. The nose horn of *Achelosaurus horneri,* or centrosaurine.3, the most recent of the Landslide Butte ceratopsians, is flattened down and attached to the snout, giving the impression that the animal is in fact hornless. There are no knobs above the eye sockets but instead deep, rough gouges. The snout is likewise pitted, as well as ridged, suggesting that it had been overlain by a boss—a bonelike protective sheath like the covering on the horns of present-day sheep. Finally, the neck shield is similar to those of centrosaurine.1 and centrosaurine.2, except that the two prominent spikes curve outward at a somewhat sharper angle.

Having found juvenile specimens in each bone bed, we were able to determine the degree of variation in skull features during maturation. Most striking, the juveniles of the three different populations are identical. Each has two shield spikes, a small, forward-pointing horn, and a slightly raised knob over the eye sockets. In other words, the three centrosaurines can be distinguished only on the basis of morphological characters that show up in adulthood. This in itself suggests a very close evolutionary relationship. (The same phenomenon can be seen in duck-billed dinosaurs. Juvenile maiasaurs, for instance, are almost indistinguishable from juvenile hypacrosaurs.) But what is even more interesting is that another centrosaurine, from the formation above the marine Bearpaw Shale—in short, after the Bearpaw Transgression—looks very much like centrosaurine.3. Called *Pachyrhinosaurus,* it's commonly found in gigantic bone beds in the Horseshoe Canyon Formation, which is what a certain portion of the Hell Creek Formation is called in Alberta. *Pachyrhinosaurus* has an extremely rough, highly raised nasal boss that extends from the front of its face backward, over its eye sockets. A pair of curved spikes jut from its neck shield.

Styracosaurus, the three new centrosaurines of Landslide Butte, and *Pachyrhinosaurus* obviously are related, yet they lived at different times—more to the point, in unmistakable stratigraphic order. To my way of thinking, this looks like a transformation sequence of one kind or another, a series of ancestor-descendant relationships—in other words, an evolutionary event.

Not everyone thinks like me, however. Strict cladists certainly don't, which is why I said that cladistics is only one of many tools I carry in my conceptual toolbox. Cladistics is inherently conservative. Though it's the best method available for determining evolutionary relatedness, it isn't good for much else. I won't burden you with the fine print of the cladistic contract. But reduced to essentials, it promises no more than to arrange taxa in terms of shared derived characters. With respect to the five horned dinosaurs, for example, a cladist would acknowledge their relatedness but without trying to describe the exact nature of their actual evolutionary relationships. Perhaps the stratigraphic data is incomplete, he'd say, obscuring the fact that the animals all lived at the same time. Or maybe we are observing nothing more than what Linnaeans would call breeds, centrosaurine.1 being the equivalent of a Great Dane and *Pachyrhinosaurus* a Chihuahua, and both related not by virtue of one having descended from the other but by virtue of both having descended from an unnamed, perhaps unknowable ancestor.

This confused me, as it may you, until I realized that cladists operate on the same assumption as the punctuated equilibrium theorists. Neither seems to believe it's possible to see evolution at work in the fossil record. The difference between the two groups is that whereas the punctuated equilibrium theorists feel justified ascribing apparent evolutionary change to this or that deus ex machina, the cladists refrain from making any assertions about the cause or direction of change, in the conviction, it seems, that about such matters no one can really know for sure. In effect, cladism is to the existence of evolution what agnosticism is to the existence of God. Here's another way of putting it: Think of individual organisms and individual characters as dots on a page. Cladists are willing to say that the dots probably represent a pattern but they don't believe there's a basis for connecting them, thereby rendering specific relationships—evolutionary relationships—visible.

My position, on the other hand, is that if there's anything we can be sure about it's evolution. Darwin's insight, that organisms descend with modification, is the fundamental fact of life on Earth. This means that in principle at least we can reconstruct every single generation that's ever existed. When I examine the fossil record, therefore, I do so in the expectation that I may be able to see evolution. Not all of the time. Not always as clearly and convincingly as I'd like to. But there's no reason to think that I'll never see it. When I'm lucky enough to reconstruct a progression of key characters like those exhibited by the five horned dinosaurs and, moreover, demonstrate, through detailed stratigraphic analysis, that it also represents a progression in time, then I feel confident in saying that the progression is in all likelihood a series of ancestor-descendant relationships—not the entire series, by any means, and not necessarily a direct one, but representative members of the same hereditary lineage.

This interpretation makes all the more sense when you consider what was taking place on the coastal plain as the seaway approached its maximum point of expansion. By the time the three centrosaurine populations appear, none of the original lowland environments exist. More precisely, the lowlands are advancing westward, reducing drier, better drained uplands to a very narrow belt, perhaps only a few miles across, squeezed against the mountains. Vegetation changes. Habitat disappears. The number of niches that any particular group of animals might occupy dwindles. There are drastically fewer adaptational opportunities. The environment, in other words, is placing an enormous amount of stress on its inhabitants, and environmental stress is one of the driving forces behind evolutionary change. In this instance it was so intense and comprehensive that entire populations were decimated. Extinctions occurred, probably lots of them. Among dinosaurs these included some lambeosaurs, *Corythosaurus,* for instance; such ceratopsians as *Chasmosaurus;* and the tyrannosaur *Albertosaurus.*

How about the dinosaurs that survived the steady loss of habitat niches, of adaptational opportunity? What impact did environmental stress have on them? To answer that question we need only consider the ceratopsian skulls of Landslide Butte. Each of the morphological characters that underwent pronounced variation between centrosaurine.1 and centrosaurine.3 is a form of sexual ornamentation. Among herding dinosaurs, horns, spikes, and shields served the same purpose they do among herding animals today. They were a means of recognition, for males in particular, to determine hierarchy, and thus leadership within the herd, as well as to attract the attention of potential mates. Females chose the reproductive partners they thought best for producing offspring, a process that selects for characters that confer an advantage in courtship and mating, in short, such display features as horns and shields. Stated succinctly, environmental stress accelerated sexual evolution among the taxa that survived the Bearpaw Transgression. As the waters rose and habitat shrank to a fraction of its former extent, sexual ornamentation ran riot with novelty. And that's nothing less than creation made visible—the tempo and

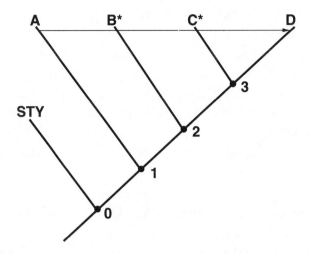

A modified cladogram to depict the evolutionary hypothesis that the
four centrosaurines are not only related by derived characters, but also
depicts, by a directional arrow, the hypothesized direction in which
change occurred. The derived characters at the branching point of the
number 0 includes having three or less spikes on each side of the
parietal shelf (STY designates the horned dinosaur *Styracosaurus*). A
character at branch number 1 is taxa A through D sharing a single
spike on either side of the parietal shield. Taxa B through D have nose
horns that protrude forward, and C and D share the character of having
extremely rough and thickened bosses over their nose and eyes.

direction of evolutionary change, the mechanisms driving it, the
consequences for particular taxa.

Different but equally influential factors would have come into play
during seaway regressions, when, in effect, environmental stress
was released. As the coastal plain expanded, habitat niches
increased in both number and variety. More and more resources
became available. Adaptational opportunities multiplied so rapidly
that animals faced far less competition from other animals, both
their own kind and others. Populations grew appreciably, and as
they did reproductive success came to depend less on sexual display

features than on characters that best equipped an animal to flourish within newly created niches, characters like modifications in the legs, enabling it to run faster or negotiate new types of terrain, or differently shaped teeth, with which it might eat novel types of vegetation. Such changes, resulting in a better "fit" between organism and environment, are examples of adaptive evolution. And judging from our excavations near the Two Medicine River, specifically, Hillside Quarry, adaptive evolution took place on the coastal plain during the Claggett regression.

Hillside Quarry, you may remember from the fieldwork chapters, represents one of the oldest layers of the Two Medicine Formation that we explored. It dates to 79.6 million years ago, precisely when the Claggett expansion came to an end and the regression began, and it is where we found *Gryposaurus latidens*. Significantly, the skull, wide batteries of teeth, and heavy leg bones of that primitive duckbill are virtually identical to those of an earlier ornithopod, the iguanodons that lived before the Colorado transgression, a group that often is defined to include the tenontosaurs of the Cloverly Formation. That in itself is a suggestive if fragmentary evolutionary connection. But there is more. *Gryposaurus* also exhibits characters shared by two taxa found in the upper layers of the Judith River Formation, that is, from sediments representing the end of the Claggett regression. Both dating from 75.4 to 74.5 million years ago, and thus having lived at the same time, the two possess, among other gryposaur features, an arched snout and a uniquely shaped pubic bone.

But they also differ in important ways. The limbs of *G. incurvimanus,* while shorter than those of *G. latidens,* are longer than those of *G. notabilis.* Similarly, the dental battery of *G. incurvimanus,* though more tightly packed than that of *G. latidens,* isn't as tightly packed as that of *G. notabilis.* Limb proportion and teeth shape, remember, are features that determine an individual's environmental fitness. Finally, judging from certain key characters, *Gryposaurus latidens* helps show how the early Cretaceous iguanodons are related to the late Cretaceous duckbills. Dental batteries,

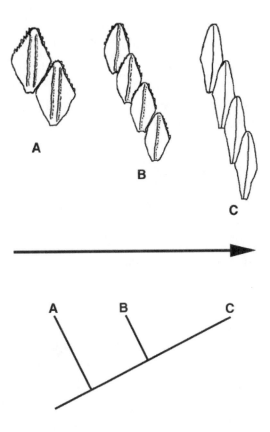

Sketch showing the teeth of *Iguanadon bernissartensis* from Belgium (A); *Gryposaurus latidens* from the lower strata of the Two Medicine Formation (B); and *Gryposaurus notabilis* from the Judith River Formation of Montana (C). The cladogram shows how the three are related, and the arrow depicts my hypothesis regarding the direction of descent.

most notably, underwent modification not only from the iguanodons to the gryposaurs, and within the gryposaurs themselves, but from the gryposaurs to the more recent hadrosaurs. Throughout the entire lineage, teeth became narrower but more numerous, arrayed in increasingly larger numbers of rows, which led to a progressive widening of the chewing surface.

Those are the dots. Here's how I connect them: When the sea

level started to rise 97 million years ago it surely created stress, and in the extreme, because the Colorado transgression eventually reached the Rockies, not merely reducing the habitat of the coastal plain but destroying it altogether. Any dinosaur that survived that profound environmental disturbance (and clearly many, if not most, didn't) were driven into the hills and mountain valleys that lay immediately west of the seaway. They became geographically isolated from one another. Likely some of them found themselves inhabiting islands. And as Ernst Mayr first pointed out, geographic isolation leads to reproductive isolation, which, under the selective pressure of new habitats, triggers adaptive changes. In time, diversity of habitat yields diversity of organisms. An incomplete and largely unexplored fossil record may prohibit us from describing in detail how the geographic isolation of the Colorado transgression affected particular lineages, but one thing seems certain: Among dinosaurs, the greatest amount of diversity followed in the wake of the highest rise in sea level. (The Colorado transgression, bear in mind, was driven by a rise in the world's oceans. All continents were being flooded at the time.) In Montana the diversification is most evident in the faunal assemblage of the lowland Judith River Formation. In all probability, *Gryposaurus latidens* is one of the links, via adaptive evolution in the Rockies, between the iguanodons of the Cloverly Formation and the primitive duckbills of the Judith River Formation.

Admittedly, the evidence that *G. latidens* emerged during a diversification of life following the release of environmental stress is circumstantial. The case for its relatives, *G. incurvimanus* and *G. notabilis*, however, is more direct. Though certainly less stressful than the Colorado transgression, the ensuing Claggett transgression significantly reduced and altered habitat, affecting local populations. Similarly, when the sea level started to drop, 79.6 million years ago, the dinosaurs of the coastal plain were presented with new adaptational opportunities—new occupations, so to speak. *Gryposaurus*, I think, represents one of the lineages that took advantage of the opportunities. Over the course of several million years, a dinosaur that had been inhabiting the upland plains— we find its earliest remains, remember, in the Two Medicine Formation—moved into the lowlands and in the process gave rise to two new daughter taxa, each adapted to slightly different niches and thus exhibiting differences in certain characters, most notably, teeth size and limb proportions.

It seems, then, that buried within the graveyards of the Two Medicine Formation were clues to not one but two modes of biological novelty, and they are complementary. When environmental stress increased, reducing habitat niches and adaptational opportunities, sexual selection occurred among single lines of evolutionary ancestors and descendants, none of which lived at the same time. During periods of stress release, by contrast, when new niches were made available and opportunities increased, the result was adaptive radiation, producing numerous related but significantly different descendants living side by side.

For the past several decades there has been an intense debate within the scientific community regarding the overall design of evolution. Does it follow a straight line or a branching course? Incredibly, duck-billed and horned dinosaurs provide the answer: Evolution does both. I've never been part of a more important paleontological discovery than this; finding eggs and babies certainly pales by comparison. In the evolutionary saga that unfolded along the coastal plain during the late Cretaceous, the dinosaurs not only show us how they came to be, they suggest how all living things, including ourselves, come to be, change, then die out. Both as individuals and as a group, their fortunes and ultimate fate echo throughout all of creation.

11

EXTINCTION AS
A WAY OF LIFE

When the Bearpaw regression began about 74 million years ago, the water didn't cease falling until it had withdrawn off the continent, never to return to the upper Great Plains. By 66 million years ago, most of what is now Montana, Wyoming, and the Dakotas was dry. A million and a half years later, the Western Interior Seaway, which had played such a crucial role in the evolution of life along the Rocky Mountain Front during the Cretaceous period, was gone, its entire contents having drained into the Gulf of Mexico. But even in its third and final retreat, the seaway continued to shape the natural history of North American dinosaurs.

Unlike the Colorado transgression, which, by isolating dinosaurs, caused them to diversify to an extent not seen before or after, the Bearpaw transgression appears to have decreased their overall diversity. Among certain lineages, I now believe, sexual evolution was greatly accelerated, but many more simply went extinct, presumably as a consequence of the intense competition that occurred as the advancing water corraled all of the remaining coastal plants and animals against the mountains, radically reducing habitat and natural resources.

To be sure, when the seaway reversed course and environmental stress started to diminish, a certain amount of diversification took place. Taking advantage of the sudden increase in enviromental opportunity, the survivors underwent adaptational changes, which increased their fitness for the novel habitat niches being created on the plain. As indicated by *Pachyrhinosaurus* and other taxa, in response to the new opportunities, the legs and teeth of ceratopsians became bigger, but there were no further changes to the horns. Among duck-billed dinosaurs, legs, too, changed; they increased in length, while the teeth became narrower and far more numerous. But diversification slowed down early in the regression, perhaps because in the wake of a perpetually retreating sea the newly exposed land tended to dry out, taking on much the same features everywhere, and thus there was less environmental variety than there might otherwise have been. Irrespective of the cause, from that point onward the dominant development among dinosaurs was rapid population growth. Instead of continuing to evolve in response to new environmental circumstances, they simply grew in numbers.

This trend is most evident among duckbills and horned dinosaurs. Based on the skeletal remains that have been recovered from the Hell Creek Formation and its equivalents in Canada, Alaska, and other western states, edmontosaurs roamed the countryside in herds comprised of tens, perhaps hundreds of thousands of animals—very much like American bison at their peak, before the arrival of Europeans. A pachyrhinosaur bone bed in Alberta contained the remains of at least a thousand individuals. *Triceratops*, the largest and most recent of the ceratopsians, traveled in massive herds as well. Curiously, we don't know where exactly *Edmontosaurus* or *Triceratops* came from—in terms of hereditary lineage, that is. The late Cretaceous sediments that might contain clues to their nearest relatives don't exist; they have been eroded away or buried. Even more curious, we don't know where exactly *Edmontosaurus* or *Triceratops* disappeared to. Nor, for that matter, do we know what became of any of the other duckbills or ceratopsians that lived in North America at the same time, between 74 million years ago and 65 million years ago. Indeed, after that juncture the fossil record contains no trace of them, not a bone, not an eggshell fragment, nothing. The dinosaurs, or what was left of them, seem to have disappeared in the geological equivalent of an instant.

You're probably acquainted with this puzzle—the apparent mass extinction that occurred at the transition between the Mesozoic and Cenozoic eras—more familiarly, between the Cretaceous period of the former and the Tertiary period of the latter. The so-called K-T boundary is in fact defined by the global die-out of the dinosaurs, or at least it was, until it was discovered that a meteor slammed into Earth at or near the boundary. If in recent years you've followed the various scientific disputes regarding the lives of dinosaurs, no doubt you've heard about this as well, because it's the basis for the most popular explanation for their demise. Even so, I don't buy it.

According to the standard version of the impact theory, an extraterrestrial body two miles in diameter and traveling at enormous speed struck Earth almost exactly 65 million years ago. The collision sent water, dirt, dust, and rock flying into the stratosphere, where it was distributed around the globe by high-altitude air currents. Like volcanic fallout, though involving a much greater volume of material, the floating debris prevented sunlight from reaching the surface of the planet. Temperatures plummeted. Plants died, followed by all of the large plant-eating animals. Those that didn't starve to death likely couldn't endure the cold climate. For a time the meat eaters may have had a field day, but soon they perished too. Among proponents of this scenario, estimates of the time that passed between the impact and its catastrophic consequences range from one to ten years. Argon dating, however, the method used to measure the age of rocks, is accurate only to within a margin of 250,000 years, which means that any specific date could be off by that much, and in either direction. The only way to measure the length of an event that transpired in less than a half million years is to examine the relative positions of various sediments, determining whether a certain stratigraphic layer is older than another, a method that offers insight but not precision.

Reservations about the supposed suddenness of the die-out aside, I think there's good reason to believe that a large meteor collided with the planet 65 million years ago. The most persuasive evidence is a thin layer of iridium found throughout the world in rocks of that age. Iridium is a heavy metal that's uncommon on Earth but often present within meteorites. It's difficult to imagine how an otherwise rare element could be universally and uniformly distributed without having fallen from the sky at the same time—from, say, an airborne layer of impact debris surrounding the entire globe. Shocked quartz, which exhibits fractures many geologists associate with impacts, is also present at the K-T boundary. Finally, a few years ago researchers discovered Chicxulub, a buried nonvolcanic crater in the Gulf of Mexico that's more than one hundred miles across. The size, structure, and composition of the crater is consistent with a two-mile-wide body hurling toward Earth from space. It seems to date from the same period.

As I see it, the weakness in the impact theory isn't the case for the impact per se but the alleged link between the impact and mass extinction at the K-T boundary. It's an imaginative proposition, but I don't believe the current evidence supports it. I hasten to add that I have no appetite for locking horns with the impact theorists, no more than I do with those who insist that all dinosaurs were cold-blooded. As I've tried to make clear in the preceding chapters, my primary interest is how dinosaurs survived, collecting stories of their individual and social lives and, where possible, assembling them into sweeping evolutionary sagas. I don't find the problem of how the dinosaurs died out to be particularly fruitful or provocative. As with the loss of individual taxa, however, or, for that matter, the issue of temperature regulation, the puzzle of mass extinction can be used to explore larger, more significant questions regarding the nature of life on Earth. In particular, I'm convinced that the current debate suffers needlessly from certain misconceptions that if exposed and corrected would not only help clarify the problem of mass extinction but shed light on evolution as a whole.

What in general is known about mass extinction? One thing we can be sure of is that the K-T event isn't unique. Fossil evidence tells us that at other times in the history of life many kinds of organisms apparently disappeared simultaneously. In fact, the largest die-out on record took place much earlier, 230 million years ago, and it marks the boundary between the Permian period of the Paleozoic era and the Triassic period of the Mesozoic era. During the Permo-Triassic extinction, upwards of 90 percent of all organisms perished. Opinion also varies regarding the factors that contributed to this catastrophe, but examining it in conjunction with the one that occurred at the K-T boundary will give us a clearer picture of what might have happened to the dinosaurs.

Before proceeding further, we would do well to remind ourselves that all of the precautions that apply to the interpretation of the fossil record also apply to the interpretation of fossil evidence for mass extinction. How can we be sure, for instance, that the record is not a distorted reflection of global events of 230 million years ago? And if it were distorted, how would we know it? Consider what is probably the most notable feature of the Permo-Triassic extinction: Although nine-tenths of all living things appear to have died out, the basic body plans that existed at the time survived. In other words, what we traditionally refer to as phyla—sponges, arthropods, vertebrates, and so on—somehow weathered the largest extinction event on Earth. All of them. Don't be fooled by the taxonomic fallacy, mistaking collective abstractions for concrete realities. I'm simply saying that at least some taxa representing each of the groups we call phyla were spared by whatever it was that killed most of the other taxa in those groups. And this is odd. One would think that an event of such magnitude would have decimated all of the members of at least one phylum. What could account for this pattern of selectivity?

Try to imagine what the planet looked like back then. The major continents were joined together in one gigantic supercontinent, Pangaea, which was centered over the equator. Although the climate wasn't as warm at the end of the Permian as it would become by the Cretaceous, it was nonetheless appreciably warmer than it is today. In all likelihood, then, the entire midsection of Pangaea, a good distance north and south of the equator, and its interior especially, supported tropical or semitropical habitats. This is important to keep in mind because most of the surviving sediments from the Permo-Triassic boundary are tropical in origin and, what's more, evolution at low latitudes differs markedly from that at high latitudes, yielding different groups of organisms. At high latitudes, where climate conditions vary greatly and often are quite harsh, plants and animals have to be very tolerant; they tend to be generalists, capable of surviving under a wide range of conditions. Since environmental niches don't vary much, any organism that does exist is likely to be the only one of its kind, or one of only a handful, none of which have changed much over time, a situation that leads to a full range of basic types—phyla, classes, orders—but relatively few taxa within each type. In Canada today, there are only three or four kinds of bear, deer, and rabbit. And if any beetles live sixty-five degrees north of the equator, you can be sure they are few in number and as tough as cockroaches.

At low latitudes, circumstances are reversed. The climate, being moderate and more or less stable, encourages the evolution of closely related specialists that fill every little niche, and there are countless, slightly different nooks and crannies in a tropical terrestrial environment. In consequence, the basic types of organism will be represented by an extremely high density of taxa. The American entomologist E. O. Wilson, for example, once identified forty-three distinct species of ant living on only one tree in the Amazon forest. In addition, low-latitude animals are highly vulnerable to temperature change, indeed, because they are so specialized, to even the slightest environmental disturbance. If the global climate shifts, they'll be the hardest hit.

And it just so happens that about 230 million years ago the seas that had occupied much of central Pangaea, contributing to its tropical climate, began to recede, leaving behind a hot, dry desert. Habitat shrank and became considerably less diverse. Thus, the Permo-Triassic mass extinction may have been massive only in the sense that it disproportionately affected a massive number of low-latitude taxa, sparing the less numerous but much hardier high-latitude taxa—and thus all of the basic body plans then in existence. The picture is further skewed because the depositional environments in which high-latitude taxa might have been preserved have vanished as well.

So, does the Permo-Triassic extinction say anything relevant to the disappearance of dinosaurs at the close of the Cretaceous period? Yes, it does. It says that massive die-outs, even die-outs that appear to have been worldwide, can occur in the absence of an extraterrestrial agent, which is important to remember as we search for parsimonious explanations. Environmental conditions at the Permo-Triassic boundary also suggest that extinctions that are thoroughgoing but confined to a narrow habitat space can look like global events. Last, after looking at different instances of widespread die-out, it becomes evident that the term "mass extinction" obscures the processes we are trying to observe because, like the terms species and formation, it doesn't resolve the processes in sufficient detail. Before we can determine the cause of a mass extinction, we had better be sure a mass extinction actually took place, and that raises a more fundamental question: What is extinction? We use the word all the time. I've used it often in these pages. But what do we mean by it? What, to be more specific, caused the extinction of, say, the passenger pigeon? Was it all the people who shot them over the years or the disease that killed the last one?

This isn't an idle question. Nowhere on Earth has a dinosaur been found in association with the K-T iridium boundary, neither on the layer nor above it. The closest any remains have been is about nine feet below it. That's where the *T. rex* we found was located, for example: under nine feet of compressed sedimentary rock, representing many more times that width in actual sediment. Since sedimentation rates vary, depending on a host of environmental factors, it's impossible to determine precisely how much time nine feet represents but it's significant, perhaps as long as 100,000 years. Those who believe that the extinction of the dinosaurs was a sudden event explain the absence of dinosaur remains at the iridium layer this way: Since the impact and its global consequences occurred rapidly—ten years or less—it's very unlikely that evidence of the event will show up in the fossil record. Sound familiar? This looks like another instance of postulating the existence of something while at the same time asserting that it can't be observed.

Apart from the difficulties this position causes for someone who'd like to back up the claim with positive evidence, there's merit, of course, in the argument that a short interval of geologic history might be missing from the fossil record. Indeed, given the vagaries of deposition and the virtual certainty of later disturbance, sometimes including complete destruction by means of erosion, we can be sure that countless ten-year periods either weren't recorded or were eventually erased, and thus any fossils they might have contained haven't been preserved. The iridium layer, however, has been preserved around the globe, and it's the result of deposition. If the ten-year period when the dinosaurs supposedly went extinct is missing, why wouldn't the only worldwide evidence for the impact that caused the extinction have been erased as well? It stands to reason, doesn't it, that some of the dinosaurs that were alive when the iridium was falling from the sky would have been preserved along with the iridium? This anomaly alone makes me wonder whether the fate of the dinosaur wasn't similar to that of the passenger pigeon. In the end, when the meteor struck Earth, there may not have been many of them around.

Attempts to sidestep this problem include examining fossils that can be found in periods prior to the impact, immediately below the iridium layer, with an eye toward detecting mortality trends—whether there was a gradual decrease in the number of dinosaurs, for instance. Most recently, Peter Sheehan, a geologist with the Milwaukee Public Museum, and his colleagues surveyed the Hell Creek Formation in eastern Montana and western North Dakota. Representing the last sediments deposited during the late Cretaceous, the formation in that part of the upper Great Plains ranges in thickness from 225 feet to 290 feet. The Sheehan group divided the column into three roughly equivalent intervals, then searched each interval for dinosaur fossils, logging, by their count, some fifteen thousand hours of fieldwork. The researchers found no significant difference in the numbers of fossils from the bottom of the column to the top, and on that basis concluded that dinosaurs didn't die out gradually during the late Cretaceous but instead remained stable until the very end, when all of them succumbed during a sudden mass extinction.

In principle, at least, the approach the Sheehan group adopted holds promise. If the extinction of the dinosaurs occurred not all at once but piecemeal, over a long stretch of time, and if a representative sample of their remains were preserved in the fossil record, then there's a good chance that a thorough survey will turn up evidence for that pattern. Difficulties arise, though, the minute one tries to define diversity, because that depends in turn on the taxonomic assumptions one makes. Going back to my point about evolution at different latitudes: Which is more diverse, one hundred species from one order, the beetles, or ten species, each of which represents a different insect order? Clearly diversity can have a different meaning, and thus degree of usefulness, under different circumstances.

Not being dinosaur specialists, Sheehan and his group identified fossils in the Hell Creek Formation only to the level that we traditionally refer to as family. When they said they found no decrease in overall diversity over time they meant that the three intervals contain roughly the same number of families. But limiting the survey to that level could mask a very different trend at the genus or species level. If the extinction of the dinosaurs was gradual or even intermittent, in all likelihood its effects would be more random than a single, abrupt, and massive event, eliminating some members of some groups but not necessarily all members of any one group. When we replicated the Sheehan study, but at a finer level of detail, that's in fact what we found—a steady decline in the number of genera from the oldest layers of the column to the youngest.

I don't mean to imply that our work resolves the issue or is in any sense the last word on the subject. I mention it only to demonstrate the pitfalls of trying to identify mortality trends in the fossil record. Imagine, for the sake of illustration, that we survey the Mesozoic era at the level of order. We would see no change at all. There are only two orders of dinosaur, the saurishia, or lizard-hipped group, and the ornithischia, or bird-hipped group. Both appeared suddenly at the beginning of the Triassic, persisted for about 150 million years, then just as suddenly vanished at the end of the Cretaceous. And, even more absurd, if we looked at the entire history of life from the standpoint of phyla we would be forced to conclude that no extinctions have taken place for the past 600 million years. Every phylum present at the beginning of the Cambrian period is represented today by living organisms.

My second objection is that the Sheehan group treated skeletal fragments as individuals that had died where the fragments lay, and there's nothing to justify such an interpretation. If anything, we should assume precisely the reverse: Any bone found by itself quite likely has been moved, maybe more than once. And determining the means of transportation and how far a bone has traveled is very difficult. More often than not, it's impossible. Besides the streamside and water hole scenarios I've already described for some of our sites there's the added complication of redeposition, which is a common geologic process. At this very moment, for instance, dinosaur bones are weathering out of late Cretaceous sediments along the banks of the upper Missouri River. Occasionally they fall from cliffs into the river. Sometimes the water rises and sweeps them downstream. Eventually the bones will be buried again, the sediments transformed into sedimentary rock, and sometime in the far future a gullible paleontogist will be led to believe that dinosaurs roamed Montana at the end of the twentieth century. Single, isolated bones don't tell us anything except to continue looking—for the real thing. In this case the real thing is an articulated or closely associated skeleton, the only proof that an extinct animal has died in place. My crew found exactly one of these during our survey of the Hell Creek Formation. Neither study, then, is reliable.

The other fossil evidence often cited in support of a mass extinction at the K-T boundary is the apparent sudden disappearance of tremendous numbers of marine invertebrates. But this evidence is questionable as well, and for two reasons. First, most of the oceanic sediments deposited during the Cretaceous have been subducted, driven into the earth along the deep ocean trenches where one tectonic plate passes below another. Since the original depositional environments have been destroyed, we don't know whether large-scale extinctions occurred in the open ocean. Indeed, we have no idea what happened there. All traces of "there" are gone. Second, the record that has survived represents marine sediments deposited as the inland seas retreated from the continents. But what might we expect to be the fate of organisms inhabiting those waters at the time? Very near the K-T boundary marine invertebrates died, all right, in untold numbers, but not because of the global consequences of an extraterrestrial collision. Their habitat simply dried up. Whatever secrets late Cretaceous marine sediments hold, and my guess is that they still hold quite a few, they probably will bear less on the issue of impact-induced mass extinction than on the role of home-grown environmental stress in the evolution of seagoing organisms. And that, I think, is a far more fertile area of research anyway.

Since I have gone to some length to raise suspicions about the impact theory of mass extinction, it's only fair that I propose an alternative explanation for the demise of the dinosaurs, one that is consistent with the geological and paleontological evidence.

The most compelling insight that the fossil record offers into the fate of the dinosaurs is that at all levels of the taxonomic hierarchy, diversity was greatest from about 80 million years ago to about 75 million years ago, during the Claggett regression and the early stages of the Bearpaw Transgression, when the environmental stress caused by the advancing sea and corresponding loss of habitat was relatively mild. North America, for example, was home to a tremendous range of hadrosaurs and lambeosaurs, ceratopsians, ankylosaurs, raptors, and tyrannosaurs, and they were everywhere, many of them in immense populations. Popular accounts still refer to the Jurassic period as the zenith of dinosaur existence, mostly because that's when the largest of them—the sauropods—reached their zenith. From the standpoint of biological variety and total numbers, however, they actually reached their peak in the late Cretaceous, about 75 million years ago.

But as I explained at the beginning of the chapter, by the time the seaway receded, many kinds of dinosaur had already gone extinct. Despite the novelty that emerged in some lineages, overall diversity was severely reduced, and once again I'm basing this assertion on fossils representing all taxonomic levels. Not only were there significantly fewer species, entire genera, families, and suborders are missing. The surviving taxa then spread out and colonized an ever-expanding and environmentally monotonous plain. Their numbers multiplied, filling gigantic habitat niches. Competition for resources was minimal. No particular taxa became geographically isolated. None of the engines of evolution, in other words, were operating anywhere near capacity.

At that juncture, several factors may have contributed to the extinction of the late Cretaceous dinosaurs. Large numbers of the same taxa, for instance, living and reproducing together, are highly susceptible to disease. Paleontologist Robert Bakker has suggested that as the inland sea retreated, dinosaurs from the eastern part of the continent mixed with those of the West, exchanging bacteria and other pathogens for which the receiving group had no inherent resistance. If that was the case, disease would have spread like flames through prairie grass. The climate was changing substantially at the time as well, becoming cooler, more arid. At no point in the history of life was the planet as warm as it was during the Mesozoic era.

Whether wholly ectothermic or, as I believe, in possession of a wide range of metabolic strategies, including mesothermy, dinosaurs would have at least suffered greatly in an appreciably cooler world. Without the ability to hibernate, the larger ones would have been at greatest risk, unable to generate enough heat to fuel movement and sustain other vital bodily processes as ambient temperatures dropped below certain levels.

Precisely where those levels might have been is anyone's guess, of course, as is the exact role any one factor played during the final days of the dinosaurs. My aim in presenting this scenario is to show that the impact theory is superfluous. Long before the K-T meteor struck Earth, all but a relatively few kinds of dinosaurs had gone extinct, and those that managed to survive were highly vulnerable. If there's any mass event that requires further investigation, it occurred ten million years before the K-T boundary and in association not with a falling star but with a rising sea. That's when most of the dinosaurs died out. If it can be said that one factor was chiefly responsible for their extinction, and I'm not yet convinced it can be said, its mark will be found in upland sediments from that period. Identifying who or what killed the last standing dinosaur, as with who or what killed the last carrier pigeon, isn't likely to explain anything of consequence. Even if it happened to be the meteor impact, which in every other way, of course, was a spectacular event, it really doesn't add much to what we know about the lives of dinosaurs.

A final thought: I detect in our desire to "explain" extinction an element of discomfort that has less to do with understanding the natural world than with making sense of the human condition. Granted, students of life strive to comprehend all sorts of events and phenomena, by identifying causes, constituents, consequences, and the like. But it seems that the attempt to explain extinction sometimes carries a special, unwarranted urgency—as if there were something odd, even unnatural about it, when in fact the disappearance of organisms is no more unusual than their appearance. Extinction is to kinds of organisms what death is to individuals. Life as it has evolved on Earth would be impossible without it. It has always occurred, it's taking place right now, and it will continue to do so, as will the evolutionary processes that produce new taxa.

To this extent, then, extinction needs no explanation. It simply is. Like death, it needs to be acknowledged and appreciated, and in the appreciation one may undergo any of a number of responses: Humility. Fear. Wonder. Especially wonder. There's nothing quite like the study of 600 million years of creation—life and death entwined—to make one take notice of the miracle that is individual existence. In our case, the miracle includes a fascinating new dimension, consciousness, which in turn gives us something that no other creature on Earth seems to possess—an acute awareness of our own mortality. Memory of what once was and foreknowledge of what will be set us apart from other organisms. Paleontology can be seen as a grand elaboration of this capacity, an attempt to bring all of natural history—that is, everything that has ever lived and died—into human awareness. And one can't long contemplate that history, with its immense and varied cast of characters, most of whom are now departed, without wondering what lies ahead for us.

12

MAN AND DINOSAUR: WHAT'S AHEAD?

Homo sapiens. Translation: knowing man. A creature that lives up to its name, and then some. Though physically indistinguishable from its immediate forebears—the late Pleistocene hominids that appeared about forty thousand years ago—*Homo sapiens* has grown clever enough to reconstruct the lives of creatures that went extinct ages ago, then to develop fictional accounts of those lives in the form of books and movies. And science and art are but two of its many notable creations. Consider agriculture, religion, language, custom and ritual, social, legal, and political institutions, economic systems, technologies, cities, the Brooklyn Dodgers, any of the aspects of contemporary existence that can be considered expressions and elaborations of culture. If *Homo sapiens* hadn't had a large brain it wouldn't have developed culture, and if it hadn't developed culture it wouldn't now call itself human. Knowing man became modern man by means of acculturation.

Some students of natural history have said that with the emergence of culture, evolution ceased for *Homo sapiens*. A buffer now exists between the selective pressures of the physical environment and the reproductive fortunes of the individual, making us less dependent for our survival on characters modified through descent than on technologies, institutions, and the like, none of which are inherited, in the strict biological sense of that term. But I think this outlook misses the point. First and foremost, and I can't say this often enough, the law of life is change through time. It's the one process we hold in common with every other organism on Earth, including the dinosaurs. Natural history is our history as well. It's where we came from; it's what made us; it's who we are. As I said earlier, any family album is a record of this fact, of change through time, evolution. What's missing from the family album are indications of the forces driving change. To capture that you have to see how human beings interact with their environment.

In doing so it will be helpful to keep in mind that evolutionary change almost always starts as a minor modification of behavior. One of the current inhabitants of one of the Galapagos Islands is an iguana that lives near the shore and feeds on algae. Evidently its kind was pushed from the interior of the island to the less populous perimeter, where it now spends a large part of its life in the water, though lacking most of the physical features possessed by other, more familiar forms of aquatic life—whales, say, or fish. In the animal called marine iguana, in short, morphology has not yet caught up with behavior. How might the two eventually converge? Any slight physical variation that confers on an individual iguana an adaptive advantage in an aquatic environment—an increase in the webbing between the toes, for instance—may increase the probability that that individual will survive and successfully reproduce, passing the trait along to the next generation.

The more general observation one can make about the marine iguana is that organisms are not always perfectly matched to their environments. This is why I part company with those who say, "The natural world is so orderly, so well designed. How could it have happened by chance?" My response to this assertion is that if nature were not well designed, how would we know? Where might we find the ideal design by which we could judge whether the world as it exists today is perfect? I believe instead that if you look closely at natural history, at the lives of particular organisms, you see that evolution isn't following a plan. Certainly it's constrained by limits—the law of gravity, for instance—but within those boundaries it does nothing but conduct experiments, small-scale, extremely slow experiments in which contingency plays a major role and the price of failure or bad luck or simple exhaustion is extinction. The truly remarkable feature of this experimentation is that, given a sufficient amount of time and a certain series of interactions between organisms and their environments, novel morphological characters and, eventually, entire new organisms emerge. And, just as important, the novelties are adaptive, they work, they persist—in the case of dinosaurs, for tens of millions of years.

Among human beings the equivalent of the marine iguana's eating habits is the staggering array of activities that fall under the heading culture. These changes, remember, despite having already transformed the face of the planet, occurred largely within the last twelve thousand years, which, from an evolutionary standpoint, is a very short period of time, too short, certainly, for morphology to have caught up with behavior. Even so, some minor effects can be seen—the apparent increase in people with poor vision, for example. With the invention of eyeglasses, individuals who would have had difficulty performing any of a thousand occupations—from hunting woolly mammoths to operating an unwieldy computer—have no trouble whatsoever, thereby increasing the likelihood that what was once a life-threatening handicap, now rendered harmless, will be inherited and thus more widely distributed.

The most pressing question, as I see it, isn't whether *Homo sapiens* is still evolving but instead whether the latest innovations in the hominid line—big brain, consciousness, culture—will truly prove adaptive in the long run, irrespective of any additional physical changes that might in time emerge. We know that these innovations aren't necessary for survival. Cockroaches seem to get along just fine without them. And nothing matches the evolutionary resilience of the lowly group of organisms that includes bacteria and algae. They have been around since the early days of life on Earth, and seem capable of surviving just about any environmental disturbance short of the complete destruction of the biosphere. Where, then, are all of our newly acquired human traits taking us? What are the advantages of moving in that direction? The disadvantages? What does it say about our chances for survival to be the only creature on Earth capable of posing these questions? Does it say anything? Could it be that consciousness is nothing more than a cruel hoax, rendering us the first organisms to foresee our own end, to bear witness to our own destruction?

Somber questions, yes, and all the more so for being unanswerable. Time will tell. Meanwhile, let's pretend that we know more than we do and speculate a bit about the phenomenon of consciousness. One thing is certain: It didn't arise fully developed or come from nowhere in particular. Everything in nature is the result of incremental modification, which means that every apparent radical transformation was preceded by a very long series of very small changes, which means in turn that many of the attributes we associate with consciousness aren't entirely unique to us. There are overlaps. We have many relatives in the animal world, and under certain circumstances some of them honor us with unforgettable lessons in relatedness.

I recall talking to a keeper at the Wild Animal Park outside San Diego about a certain female chimpanzee who always sat by herself, apart from the other animals, an outcast. So disliked was she that the dominant male attacked her often, and viciously, seemingly without provocation. Why, I asked, was that particular adult female being persecuted? Several years before, it seems, she had lived at the San Diego Zoo, where, for some reason, she mistreated another adult female and her young, one of which was a singularly robust male. When that male grew up, he assumed control of the pack and when the pack was transferred to the Wild Animal Park, where environmental conditions were less domestic, he reverted to typical chimpanzee behavior, taking revenge upon the tormentor of his youth. In other words, he remembered the actions of the older female from years past and he punished her for it, not once, but repeatedly. Although more difficult to document, chimpanzees may plan ahead as well, though in a very primitive way—the unforgiving male laying in wait for the beleaguered female, for instance. That suggests some degree of foresight.

In human beings, by comparison, both of these capabilities have undergone an enormous amount of refinement. Take our obsession with looking into the future. We didn't stop at planning for tomorrow. We started wondering about the day after tomorrow, then the one after that, reaching ever further ahead, and that made us anxious. Not long afterward we discovered a Supreme Being who offered assurance that our future is in good hands. Now, there are lots of ways to talk about the relationship human beings have formed with supposed higher orders of existence, but the word most commonly used in such discussions is soul, an innermost self that stands apart from everything else as a distinct entity. Soul is a word that rarely surfaces in scientific discussions, and that's understandable, but I think there's a way to think about it that reconciles the interests of science and religion, at least tentatively.

Ask yourself this question: Of all of the attributes that distinguish human beings—the nickname we have given to ourselves—from Homo sapiens, what's the most definitive, the one without which all the others would not exist? Surely it's that point in the development of consciousness when the idea of the soul was born. You could say that the creation of human beings began with the inception of the soul, and therefore that that event separates us from other organisms, because this particular character is unique to us. Human beings came into existence the moment Homo sapiens began worrying about the distant future and formed a relationship with a Supreme Being in whose hands it entrusted its fate. If you were in a Linnaean frame of mind, you might say further that at that juncture in evolution a new type of biological creature emerged. In other words, human beings may not be the pinnacle of primate existence but instead the most primitive version of what's to come. I'm not asking you to believe this. Just play with it, as a way of thinking about the future of mankind and the dilemmas we face. Whatever conclusion you may come to about the soul, it's an idea that's had a profound impact on human history, and for that reason alone it cannot be ignored.

Another dimension of sapient existence you might want to consider is specialization and its implications for survival. The first human beings to arrive in North America, like the first organisms in any new habitat space, were generalists, capable of performing any of a wide range of occupations that might be required to stay alive under an equally wide range of environmental circumstances. They secured and prepared food, built shelter, made clothes, tended to illness and injury, domesticated animals. If no one person possessed all of these skills, the skills were nonetheless well represented in the small bands and tribes in which the people lived. But as time went by and civilizations developed, and especially since the Industrial Revolution, the environment became partitioned into increasingly smaller and more diversified niches. One can now survive as a paleontologist, for instance, without knowing how to grow corn or rice, dig a well, weave a sweater, or do much of anything else, because there are legions of specialists who themselves do nothing but farm, build and repair things, sew, take out the garbage. In midtown Manhattan, until very recently, there was a shopkeeper whose sole occupation was repairing zippers. For more than fifty years that's all he did, and there was enough call for his highly specialized craft that he made a good living at it.

Such situations are commonplace in large cities, because that's where specialism is most extreme and, as everyone knows, humanity seems hell-bent on urbanization, gathering together in increasingly larger collectives that now number in the millions, even, in a few cases, in the tens of millions. From one perspective, places like New York City and Mexico City and New Delhi can be seen as monuments to the ingenuity of human beings, their seemingly limitless capacity for innovation. From another perspective, it all seems rather precarious. Think of the high susceptibility of low-latitude organisms, the overwhelming plenitude of insects in the Amazon rain forest, for instance. Such richness of biological expression is surely one of the marvels of the natural world. But that marvel is sustained at some risk. All that's needed to destroy it is a small environmental disturbance. Though the parallel is far from exact, the marvelous diversity of modern civilization also comes with an inherent liability. Should the world's industrial economies collapse all at once, say, in response to a global decline in fuel production, cities will be especially vulnerable. Far removed from sources of food and other basic requirements, and lacking a workable range of practical skills, urban residents will be thrown into chaos. *Homo sapiens* would probably survive such a global catastrophe, but there may come a time when human beings would not, and the process of civilization would have to begin anew.

Without going into the history of religion and law, it also seems clear that when the soul came into being, whether as a concept or something more substantial, a moral dimension was added to human existence, and nowhere are the complications of consciousness more evident than in moral debates about the issue of extinction, not our own but that of other organisms with which we share the planet. Nature, as I argued earlier, is not following or fulfilling a plan. There's no master blueprint for evolution that would give us the grounds to say, "Well, the loss of that species was a big mistake." There are patterns in nature but no mistakes, no right or wrong direction, no preordained destinations from which we might be deterred. Most of life seems to understand this, more accurately, not to give it any thought. No one cried over the demise of the dinosaurs. If that event were to occur today, however, tears surely would fall. Human tears. Because, unlike nature as a whole, we make plans. We make judgments about deviations from our plans. And we know what no other creature knows—that sometimes we are to blame for the deaths of others.

I don't see any way out of this dilemma. We'll probably argue over our impact on other species and what to do about it for the rest of our days on Earth. And we'll be unable to free ourselves of the paradox of settling disagreements about the best interests of the natural world in thoroughly man-made settings—Congress, the federal courts, and so on. But meanwhile I think we should get used to the idea that the organisms that can put up with us—deer, coyotes, possums, cockroaches—are the most likely to endure, regardless of the heroic efforts we may make on behalf of others. I'm not saying that such efforts are necessarily misguided or doomed to failure, only that the overall trends of human history, increased industrialization and urbanization, coupled with continued population growth, show no sign of slowing down. This is placing a tremendous amount of environmental stress upon other plants and animals, which in turn is selecting for characters compatible with the rising tide of civilization, a process not unlike the flooding of the continents during the Cretaceous period. The difference, of course, and it's all the difference in the world, is that inland seas are without consciousness and thus untroubled by a sense of responsibility. That, it seems, is a burden we alone carry. In exchange for memory and foresight we gave up our innocence. An old story, surely, retold here in somewhat different terms. According to the evolutionary version, however, there's no going back, no way to restore the world to a previous state. Time travels in one direction. And the only escape from time is death.

Does the foregoing seem a little far afield for a dinosaur paleontologist? It doesn't to me. In one sense or another I have wandered all of my life, when not actually walking through barren hills and rocky washes in search of fossils, then letting my imagination run free. Even if this or that particular expedition fails to turn up anything of value the exercise alone keeps one limber, in body and mind, and alive to fresh possibilities, without which existence would be a pretty joyless affair. Besides, generating ideas is a different activity from testing them against the evidence, and every idea, regardless of origin, should be put to the test. That's the credo of scientists, at any rate.

In practice the second part of this process is more dynamic than it might sound. For one thing, the evidence in paleontology is always changing—expanding, being refined, sometimes undergoing complete reinterpretation. An idea that once seemed to contradict the available data is resurrected when it's learned that the data was skewed. For another, our means for collecting evidence also changes, creating new avenues of research, provoking altogether novel ideas. The heroic version of science would have us believe that it progresses under the force of brilliant, probing minds boldly interrogating the mysteries of the universe—in short, by means of direct confrontations between human beings and nature, consciousness trained like a laser on this or that puzzle. And surely there's some truth to this. But just as often, advances in scientific understanding follow advances in the tools and procedures of inquiry. Astronomy and molecular biology provide the most obvious examples. Imagine the status of these fields today if the telescope and microscope, to say nothing of the more sophisticated devices that succeeded them, had never been invented.

A similar claim might one day be made about paleontology. Using CT scans, computerized imaging programs, microscopy, and other advanced methods, we are starting to extract heretofore unheard-of kinds of information from fossils. An entire vista of discovery, micropaleontology, has opened up in recent years. Although this technological revolution is in its infancy—the most useful benefits so far coming from histologic studies like those I described in chapter 8—it's already showing great promise, convincing me that the bones paleontologists and others have been collecting for the past 150 years have a great many more stories to tell about the lives of dinosaurs. All we need do is learn how to listen. And that's precisely what we're attempting at our laboratory at the Museum of the Rockies, where the new techniques are being tested on real-world problems. Here are three recent projects:

Kristi Curry, now a graduate student at the State University of

New York, Stony Brook, has prepared histologic profiles of juvenile apatosaur forelimbs found at Mother's Day Site, comparing them with similar bones recovered from an excavation in Colorado. Her primary aim is to gain insight into sauropod growth. From the standpoint of physiology, the sauropods are the most baffling of all of the dinosaurs. They were the largest known land animals, weighing up to fifty tons, yet their heads were tiny, their mouths even more so. To get an idea of the proportions—rather, disproportions—picture your head reduced to the size of a peanut. How they managed to eat enough to survive from one day to the next is a mystery. But they did more than that. One of the most successful four-legged creatures ever, the sauropods lived for more than 100 million years. So far, Kristi has found densities of vascularization and tissue patterns consistent with cows and other ungulates, suggesting that apatosaurs grew rapidly until they were half grown, or about thirty feet long. But she has also turned up evidence of variation in growth rate in different bones of the same animal, which tells us that temperature regulation in dinosaurs may be even more complicated than we had thought.

In another effort to make fossils speak in new ways, post-graduate student Mary Schweitzer has been trying to extract DNA from the bones of *T. rex*. Originally, like Kristi, she had intended to thin-section the bones and conduct a histologic investigation. But under the microscope there appeared to be blood cells preserved within the bone tissue. Mary conducted a number of tests in an attempt to rule out the possibility that what she'd discovered were in fact blood cells. The tests instead confirmed her initial interpretation. Then, using certain chemical processes, she tried to isolate the cells. That didn't work. So next she tried, with some success, to recover proteins, including DNA and collagen, the chief constituent of skin, ligament, and bone. DNA, of course, could help determine hereditary relatedness. Complete sequences of numerous animals would be required, however, so that goal is a long way off. Collagen, too, varies from one individual to the next and thus could be used as a kind of chemical fingerprint. Regarding either prospect, however, the work Mary is doing represents one of the first small steps in what may prove to be a tremendously fruitful area of laboratory research.

My final example involves computer graphics. You'll recall that to be able to conduct our recent histologic studies of baby hypacrosaur and maiasaur bones, my wife, Celeste, and I first had to simulate near-term embryonic skeletons and position them within reproductions of their eggs, so as to determine how large the hatchlings were at birth—in other words, to make sure that the bones we were analyzing actually belonged to babies. The software for developing and manipulating three-dimensional figures has been around for some time now, and we've been experimenting with other applications. But we have also begun to explore the uses of so-called morphing programs, the most promising of which enables us to visualize how a particular dinosaur grew from infancy to adulthood.

One of the reasons this has proved helpful is that dinosaurs didn't mature linearly; the adult skull, for example, isn't merely a blown-up version of a juvenile one. Like the skulls of birds and mammals but unlike those of reptiles, the skulls of dinosaurs changed shape during development. In general they start out with long foreheads and big eyes and end up with sloping foreheads and eyes that are proportionately smaller. Imagine how odd it would be if a robin chick or a lion cub retained its baby features throughout life. Indeed, it's the peculiarity of that fictional condition, and our reflexive response to it, that explains the universal appeal of such figures as Mickey Mouse and items like teddy bears. I'll come back to this very shortly.

In our collection at the museum we now have hypacrosaur skeletons representing five different stages of growth. We made drawings of the fossils, and entered the two-dimensional images into the computer. The morphing program then averaged the five pictures and filled in the intervals between each one, giving us a continuous depiction of dinosaur growth—a little movie, so to speak. Unlike *Jurassic Park* or *The Lost World,* however, our film repre-

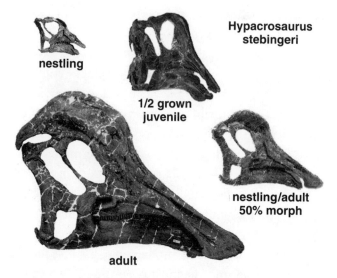

Hypacrosaurus stebingeri

nestling

1/2 grown juvenile

nestling/adult 50% morph

adult

Skulls of *Hypacrosaurus stebingeri* showing its nonlinear growth. The 50 percent morph is a linear result of morphing the nestling with the adult. The half-grown juvenile shows the actual configuration of the skull. Note that the snout hasn't elongated, and the nasal crest hasn't developed. The half-grown *Hypacrosaurus* retains its juvenile characteristics.

sents a digitized version of the real thing—skeletons. It is the best approximation yet of someone actually having been present with a camera in a hypacrosaur rookery 74 million years ago. Watching the transformation that occurs during development is an unforgettable experience. (The hypacrosaur on the book jacket is represented at about the age when it would have started to assume more adult characteristics.) As the skull matures the snout grows longer and flattens out, like a duck's bill. The eyes recede and become less prominent. Along the forehead the crest expands upward, eventually forming, in the case of the adult male, a high, narrow ridge, a sort of bony crown turned upright, in the shape of a very exaggerated mohawk.

This merely hints at what we hope to accomplish in the coming years. As we perfect the morphing software, expanding to three-dimensional programs and increasing the overall resolution of the imagery, we will assemble a library of growth series, which can then be used as a basis for identifying bones (those of different baby duckbills are especially difficult to tell apart), constructing simulations of stages for which we have no representative fossil remains, hunting for evidence of telling anatomical structures (maybe nasal turbinates turn up only at certain stages of development in certain dinosaurs), and in general exploring the relationship between developmental morphology and other aspects of dinosaur life. In the case of *Hypacrosaurus*, it seems likely that one of the reasons the young did not possess full-size crests is that it would have disrupted the social hierarchy and communication processes within the herd. The crest was one of the male adult's display features, for intimidating rivals and attracting mates. But it was hollow as well, and connected to the nasal passages, which means that it was probably used to trumpet sounds during courtship, when danger loomed, and so on. Having heedless juveniles in the herd that were capable of making these sounds would only have confused matters.

Judging from the growth of the skull, in particular, *Hypacrosaurus* also appears to have taken full advantage of a survival strategy that's common in warm-blooded animals—the retention of baby features. Hypacrosaur young kept their youthful appearance until they were about half grown, after the point when, according to our estimates, they were capable of leaving the nesting grounds. A number of studies with human children have shown that the longer one retains baby characteristics, the longer one will be cared for. Such features trigger instinctual parenting responses, even in other children, which probably explains why when a child brings home a stray, more often than not it's a puppy or a kitten, not an adult animal. More generally, if you happened to be a baby that could not care for itself, it would be to your advantage to have a face that melts the hearts of grown-ups. Which is exactly the scenario I see for *Hypacrosaurus, Maiasaura,* the Milk River lambeosaur, and many other dinosaurs. What purpose would it serve to retain baby features if from the moment you were hatched the herd treated you like an adult?

*   *   *

With our arrival at the first, experimental morphing images we've completed a circle. We started by discussing the role imagination plays in our attempts, both scientific and not-so-scientific, to reconstruct the lives of dinosaurs. We then traveled into the field to collect more fossils and more information, subjecting our finds to further study in the laboratory, extracting secrets from bones that only a short while ago we would not have imagined they possessed. The guiding ambition behind all of the work we've done since leaving Egg Mountain was to create a fuller, more forceful vision of the dinosaurs, only this time as characters in an epic drama, an evolutionary saga that spans millions of years. In all instances, though, the interplay between fact and imagination never ceased.

With dinosaurs, it can never be otherwise. The stories will always be incomplete, ambiguous, under revision, and we'll always want to fill in the missing pieces. Because we can't help it. Seeing what's not really there—remembering and anticipating—is one of the things we do best. We are, I think, a perfect match, *Homo sapiens* and Dinosauria. Knowing man meets the partially known but unmistakably real, a factual footing for an ever-restless imagination. Wherever fate leads us, then, doubtless we'll continue to turn around now and again, glancing backward, taking the measure of who we are against everything that ever was. And once upon a time the dinosaur was. It actually was. That's where we began. It's where we'll begin again.

AFTERWORD

WHAT'S A DINOSAUR WORTH?

When, in the early 1870s, P. T. Barnum took his three-ring circus on the road, he brought his usual assortment of real curiosities and wondrous fakes, including Esau, the Bearded Boy; a giraffe, which very few Americans had witnessed firsthand; and the Cardiff Giant, a ten-foot statue that he billed as the petrified remains of a prehistoric man. Barnum's show was immensely popular with youngsters, and none more so than Frank Brown, a teenager living in Carbondale, Kansas. Early in February 1873, Frank could think of little else but the "Great Traveling World's Fair," due to arrive in town only a few days later. Spellbound by the prospect of seeing for the first time what he had long been hearing about, Frank even suggested to his parents that his new baby brother, born on the twelfth, be called Barnum, and his parents, at a loss for a more suitable name, consented. What role this accidental connection actually played in the forging of Barnum Brown's personality is impossible to tell, but it's nonetheless a fact that the youngster grew up to be a paleontologist interested less in sound scientific research than crowd-pleasing showmanship.

Somehow Brown found dinosaurs just about everywhere he looked, and like his namesake, he assembled an extensive menagerie of exotic creatures, more precisely, their skeletons, most of which now rest within the collection at the American Museum of Natural History. Unfortunately, apart from what they reveal about anatomy, the specimens are largely useless—because Brown, for all his success at collecting bones, was careless about collecting information. He simply didn't consider it important to describe where he unearthed his specimens. Nor, evidently, did anyone back at the museum. Regarding the location of the *Gryposaurus* skeleton from the Two Medicine River area, Brown had only this to say: "15 miles southwest of Cut Bank." So it was fortuitous that we ran into Tom Harwood, the man who had led the early fossil hunter to the specimen. In fairness, it should be said that Brown, though extreme in his disregard of data, was a creature of his time, an era of paleontology when merely excavating and assembling a skeleton was a great accomplishment. Very quickly, however, it became clear that skeletons have much to tell about the lives of dinosaurs, but only if the contexts in which they are found are preserved and documented. Brown's brand of dig-and-run collecting fell out of favor.

Until recently. Now, the secrets the bones hold are being sacrificed not for the benefit of public shows but for the sake of private sales, a far more troubling development that's already caused the loss of a great deal of scientific information, to say nothing of the damage it's done to the traditionally genial relationships between professional and amateur collectors and between collectors of all kinds and landowners. Paleontology is one of the few remaining sciences in which amateurs still make contributions. Equipped with a shovel, pick, and a little knowledge, almost anyone can mount her own search for fossils, which is one of the reasons the field is so attractive to the public. But as the enthusiasm for dinosaurs has spread and the value of dinosaur bones has skyrocketed, so have trespassing and looting. Some landowners have responded by closing their property to all collectors, others by demanding financial compensation, which is the same as closing it to all but well-funded commercial interests. Consequently, since leaving the Blackfeet Indian Reservation I've been forced to restrict most of my research to land held by the states and the federal government. I'm luckier than many in this regard, because I live in Montana, which contains vast expanses of such holdings. But commercial interests are now targeting them as well, precipitating bitter conflicts and, more ominous, fueling efforts to rewrite the laws governing collection on public lands.

Probably the best-known controversy involved "Sue," a *Tyrannosaurus rex* skeleton whose namesake discovered her in 1990 on the Cheyenne River Sioux Indian Reservation, in South Dakota. To appreciate what was at stake you need to know that there are three types of land on reservations, each controlled by a different law enforcement agency and subject to different rules regarding access. Tribal lands are owned wholly by the tribe and fall within the jurisdiction of tribal police. Only the tribal council can grant permission to collect fossils on tribal land. Trust land, by contrast, is held by the U.S. government for the use of the tribe, tax free, the laws thereon enforced by the FBI. Collectors' access is governed by the Bureau of Indian Affairs, which issues an antiquities permit to those it approves. The third type of reservation property is deeded land, which is often owned by non-Indians. Laws in these areas usually are enforced by the local county sheriff. To gain access one need only secure the owner's permission. It was while excavating duck-billed dinosaurs on deeded land near the edge of the Sioux Reservation that workers with the Black Hills Institute (BHI), a private collection company, came across the remains of Sue, and that's when the trouble started.

I should say, where the trouble started, because the tyrannosaur wasn't found in the duckbill quarry but on nearby trust land.

Instead of requesting permission from the Bureau of Indian Affairs, as required by law, BHI paid the trustee five thousand dollars for the specimen, then removed it from the ground. Soon the tribal council learned about the transaction and, concerned about people collecting on the reservation without their approval, they asked me to examine the site where the dinosaur had been excavated and to recommend a course of action. Not only was it clear that a large skeleton had been dug up, there was unmistakable evidence that someone had been digging at several other unauthorized sites as well. I told the tribal lawyers that since the specimen had been found on federal land, its primary owner is the U.S. government, but in my estimation, since the property in question is held in trust for the Cheyenne River Sioux, that the tyrannosaur was theirs to do with as they saw fit. Recognizing that the tribe lacked the necessary curatorial resources, however, I further advised that they allow BHI to prepare and display the skeleton. Under the proposed agreement, which the Sioux lawyers accepted, BHI would have been allowed to charge admission and to sell casts of the bones, though not the bones themselves. But BHI rejected the offer, demanding sole ownership. The FBI responded by confiscating the tyrannosaur. Recently a federal court ruled that the tyrannosaur belongs to the original trustee, who in turn consigned it to Sotheby's of New York, where it will be auctioned—very possibly to an overseas buyer.

Sometimes conflicts over dinosaurs found on public lands have happier endings—at least as far as the public is concerned. One of the country's most prolific sauropod sites is near the Bighorn Mountains of Wyoming, east of Yellowstone Park, on a ranch owned by the Howe family. Discovered in the early 1930s by, you guessed it, Barnum Brown, Howe Quarry is a late Jurassic deposit which holds thousands upon thousands of sauropod bones, even today, after Brown's extensive digging (he shipped a boxcar full of specimens back to the American Museum) and numerous other collectors' digs. Toward the end of the summer of 1991, a commercial firm from Switzerland was working at the site when, as had happened on the Sioux Reservation, members of the group began exploring nearby public land without first getting permission to do so, in this instance, from the Bureau of Land Management. Soon they made a significant discovery—a virtually complete skeleton of *Allosaurus,* a theropod from the late Jurassic that grew to thirty-six feet in length and weighed up to two tons.

Excavating such a large specimen, however, would require not only heavy equipment but the building of a road on which it could be transported to the site. So that is exactly what the Swiss collectors set about doing—until two BLM agents, who happened to be flying over the area one day, spotted the activity. Under prevailing statutes, the BLM could've fined the Swiss collectors but the agency chose leniency instead, seizing the specimen and warning the firm to stay off public land. Unexpectedly in possession of a large dinosaur skeleton that they very much wanted excavated before looters learned of the site, the BLM then turned for help to the Wyoming Geological Museum, the Museum of the Rockies, and the Royal Tyrrell Museum of Paleontology. They also contacted the Smithsonian Institution, where, strictly speaking, anything of scientific, historic, or cultural value found on federal land is supposed to be deposited. The Geological Museum said that it lacked the resources to mount an excavation. The Smithsonian wanted the specimen but couldn't field a crew until the following summer.

So the BLM asked me what the Museum of the Rockies might be able to do in the meantime. "I can put a crew on the site within forty-eight hours," I said, wondering if, that late in the season, the weather would remain favorable. With the Smithsonian's blessing, the BLM permitted me to excavate the dinosaur. The day Bob Harmon and his co-workers removed the last jacketed section from the ground, snow started to fall. Now safely housed at the museum in Bozeman, the allosaur has been completely prepared. We plan to mount an exhibition cast of the skeleton, along with one of the *Tyrannosaurus rex* we excavated in eastern Montana, in the summer of 1997. In turning the skeleton over to the museum, the BLM ensured that it would stay within the Rocky Mountain region, where it was found, and under circumstances that would make it accessible to the public.

Confrontations like these, which are increasing throughout the West, have led to efforts on the part of the commercial collectors to get federal laws enacted that would allow anyone, regardless of the size of their operation, to remove fossils from public lands for the purpose of selling them. But their argument does not stand up to scrutiny. Fossils are not natural resources like oil, timber, and minerals, that virtually everyone requires to survive, which is the reason drilling, logging, and mining are allowed on public lands. Fossils are instead scientific resources; their primary value is educational. This is especially true of vertebrate fossils, by virtue of their rarity and their potential for shedding light on the history of life. Under current statutes, vertebrate fossils found on public lands belong to everyone. They are considered national treasures, a public trust placed in the care of public institutions, like the Smithsonian and the Museum of the Rockies. The legislative amendments the commercial collectors support would permit them to remove these treasures from the public domain and use them for their own private purposes, in other words, to turn a resource that now exists for the eternal educational benefit of everyone, into one that exists for the short-term financial benefit of a few.

To obscure the true nature of this transfer, commercial collectors have lately tried to recast their campaign in terms of individual freedom, an increasingly common tactic among those who dispute federal and state authority. In this view, by prohibiting the removal of vertebrate fossils from public lands, the federal government is trampling on the rights of Americans, though precisely which rights are at stake is never made clear. Capitalizing on antigovernment sentiment in the West, they have tried to rally support among amateur collectors and rock hounds, even accusing the scientific community of conspiring with the federal government to deny them access to all fossils on state and federal holdings, which isn't true either. The issue is not governmental intrusion into the lives of private citizens; it is whether the government is going to protect and maintain a long-standing public trust. If the commercial collectors get their way, that trust will be theirs to do with as they see fit, and you can be sure that the public interest will be the last thing on their minds.

If I seem aggravated, your impression is correct. But if what I've said has led you to believe that my primary concern is ownership, then I have more explaining to do. For better or worse, dinosaur bones have become expensive commodities, in some cases fetching sums of money that would have been unimaginable only twenty years ago. Today, for example, a single well-preserved *Tyrannosaurus rex* tooth may cost as much as several thousand dollars; a complete skeleton could fetch more than a million dollars. Tomorrow the prices could be higher, especially in the case of species like *T. rex,* of which there are about two dozen skeletons in the entire world. Under these economic circumstances, arguments over ownership are inevitable. By the same token, if the market for dinosaurs were to collapse suddenly, an extremely unlikely prospect, there would be few commercial collectors who bothered to campaign for the rights of the weekend collector in his four-wheel-drive pickup. Follow the money, as the saying goes, and you'll discover what's really motivating the disputes, along with who stands to benefit most from a change in federal policy.

But the unavoidable disputes over ownership tend to distract attention from a more fundamental concern, one that long predates the current controversy, going back as far as the haphazard excavations of Barnum Brown and other early paleontologists. That concern is how best to preserve the scientific value of fossils. To grasp what I mean by scientific value, imagine a spectrum. At one extreme is the skull of a duck-billed dinosaur. Apart from being able to identify the species to which the skull belongs—say, *Hypacrosaurus stebingeri*—we know nothing else about it. On the opposite end of the spectrum is another *Hypacrosaurus* skull, virtually identical to the first, but it comes with a highly detailed, verifiable story. We know the exact location, age, and composition of the sediments in which the skull was found. We know their structure and orientation as well. We also know the same things about other fossils deposited near the skull, including, if present, the rest of the skeleton to which it belongs. We know further how and when the dinosaur died, if it was preyed upon or moved after death, whether it was alone or part of a group and, if so, what kind of group. Placed side by side, the two skulls appear indistinguishable, yet the first has lost almost all of its scientific value whereas the second retains about as much as one could expect of an object seventy-odd million years old.

What accounts for the difference? Information, of course. To be specific, information about the specimen's geological context. A dinosaur bone out of context is just that—a bone. In context, however, it serves as a window onto the world the dinosaur inhabited, as well as its history and fate. And when the bone, along with the accompanying scientific information, is housed in a museum or other public institution, it's a window paleontologists can return to again and again, sometimes to see things that had been overlooked before. Some commercial collectors are more conscientious than others; they understand the importance of documenting the context in which fossils are found and do their best to collect information before it's erased forever. In the collegial spirit that characterizes nonprofit scientific research, they'll even share their data with professional paleontologists. I've enjoyed just such a relationship with a firm called Canada Fossils Limited. But data collection is difficult and, for a commercial company paying labor costs, often prohibitively expensive.

For one thing, the excavation itself is a painstaking process that requires an enormous amount of time and patience. Usually working beneath a blazing sun in desertlike locales, we remove tons of rock, one spoonful at a time, using nothing but hand chisels and whisk brooms, while carefully mapping, extracting, and cataloging every shard of bone, shell, or plant fossil we encounter as we dig. For another thing, proper scientific documentation also requires that at least one person with advanced training in geology be present at the site. Many commercial collectors operate on the basis of hearsay and general information; they know, for instance, that certain kinds of dinosaurs have been found in red sandstone beds in a certain region. But they cannot conduct the sophisticated analyses that enable us to identify exactly which kind of depositional environment the sediments represent, an upland water hole, say, or a lowland delta, or the geological events responsible for the formation of the environments. Finally, and this is a point that's little appreciated outside the paleontological community, from a scientific standpoint often the most important specimens aren't museum-quality skeletons but the countless smaller, usually damaged and almost always incomplete bones that taken in the aggregate can be used for comparative purposes—precisely the fossils, by the way, that enabled us to reconstruct the evolutionary saga described in this book. In practical terms, this means that frequently the specimens with the greatest scientific value possess little or no financial value. Commercial collectors have nothing to gain by preserving and collecting them. In fact, they are viewed as impediments, to be destroyed, like the surrounding rock matrix, to get at such highly prized objects as large skulls.

The engine that drives commercial collection, reserving its largest rewards for those who don't take the trouble to conduct scientifically rigorous excavations, is, of course, the marketplace itself. As long as there are museums, institutions, and individuals, in the U.S. and elsewhere, willing to pay extremely high prices to own or exhibit particular specimens, some collectors will go to almost any length to meet the demand. Imagine that a certain museum—let's call it the P. T. Barnum Museum of Questionable Curiosities—announces that it will pay $750,000 for an intact ceratopsian skeleton. The only accompanying information the museum requires is the dinosaur's name, the formation in which it was found, and its age.

Harry Quickbuck, of Harry's House of Fossils, a commercial company, happens to have a partial ceratopsian skeleton in storage. But never one to pass up an opportunity, especially one as lucrative as this, Quickbuck instructs his fieldworkers to find the parts needed to complete the specimen. The crew collects no paleontological or geological data. It constructs no maps of the dig. The bones may or may not represent the same species. But that doesn't stop Quickbuck. In fact, he prefers it this way, because he doesn't intend to reveal the real source of the additional parts. He assembles a hybrid skeleton and sells it to the P. T. Barnum Museum, which puts it on display for an unwitting public. Sometime later, Dr. Rightway visits the museum to study its collection. An astute dinosaur paleontologist, he detects small but crucial anomalies in some of the bones of Quickbuck's specimen. Operating on the assumption that the skeleton was found just as it's displayed, Dr. Rightway is forced to conclude that he has identified a previously unknown species, then publishes a paper describing the new ceratopsian. A fake enters the scientific literature undetected.

I offer this tale to illustrate my point about ownership. As I've already made clear, I endorse the present arrangement, under which vertebrate fossils located on public land belong to the public and when dug up are held in trust for the public—forever. In fact, I'd like to see the policy strengthened, specifically, by substantially increasing the fines brought against those who remove vertebrate fossils without a permit. Compared to what a commercial collector stands to earn from the sale of such fossils, the fines are negligible, encouraging collectors to treat them as one of the costs of doing business. But in taking this stance I'm speaking primarily as a citizen, one who's concerned about the fate of public resources. Speaking as a paleontologist, on the other hand, my attention shifts from the ownership of fossils to the protection of information. My first concern isn't that someone might make a profit from the sale of public property but instead that when the desire to make money is paramount there's no incentive for conducting scientifically sound excavations.

Most professional paleontologists follow a strict set of rules regarding how data is to be collected, preserved, and made public. They have developed a common vocabulary for describing what they do and the specimens they find; they adhere to universal guide-lines regarding the preparation and preservation of specimens; they're required to make their data and specimens available to other scientists so that their claims can be verified—and all of these requirements, taken together, guarantee that what I described in the P. T. Barnum Museum example does not happen. Sure, glitches are sometimes introduced, an error here, a misinterpretation there, but as I explained when I introduced the procedure called the Null Hypothesis, paleontology is self-correcting. Indeed, that's what drives the entire scientific enterprise: In the light of new evidence and more informed interpretations, we continually refine our pic-tures of the world. If one of my colleagues publishes a paper in which he asserts that he found so-and-so dinosaur in such-and-such sediments, I can count on what he says, even if I disagree with his conclusions, because he plays by the same rules that I do, that all paleontologists do. Here, then, in a nutshell, is why I worry about dinosaur collection being placed at the mercy of market forces: Under conditions in which the sole motive is making money, not only will priceless information be lost, and lost forever, but that which does survive will be no more trustworthy than Harry Quickbuck's ceratopsian or, for that matter, P. T. Barnum's Cardiff Giant statue. Science will be displaced by a sideshow.

Finally, I'd like to try to correct a misrepresentation that some commercial collectors have promoted in recent years. In an effort to drum up support for their cause they have tried to convince amateur collectors that professional paleontologists hold them in low esteem or, worse, would like to see them go away altogether. I can't speak for the entire scientific community on this point, but the paleontol-ogists I know personally, those with whom I've worked over the years, do not share this view. We collaborate with amateurs all the time, and we enjoy the collaboration. Because paleontological research is notoriously underfunded, nonprofessional volunteers form the backbone of many digs. Equally important, by virtue of their numbers and the ground they can cover, as well as the serendipitous nature of fieldwork, to say nothing of the high levels of skill and experience some of them bring to the task, amateurs routinely make significant finds. In both respects, I've benefited greatly throughout my professional career. So have my colleagues.

On a more personal note, for many years I was an amateur collector. If certain aspects of my life had gone differently, I'd still be one. I'd be traipsing through the badlands of Montana searching for dinosaur fossils, motivated by nothing more than the desire to witness something I hadn't seen before—to be surprised—which, come to think of it, is the same thing that motivates me today. This is why I believe that, despite some obvious but superficial differences, the weekend collector and the seasoned paleontologist form a natural alliance, because their ambitions are the same: to experience the joy of discovery, thereby increasing what we know about the world in which we live. For both groups the study of dinosaurs is a labor of love, but it will remain so only if the vertebrate fossils on public lands are protected from commercialization, which will allow us, amateur and professional alike, to collect and preserve the data needed to tell dinosaur stories in full, for the benefit of everyone.